EXPERIENTIA SUPPLEMENTUM 20

New Concepts
in Air Pollution Research

Interdisciplinary Contributions
by an International Group of 20 Young Scientists

Edited by

Jan-Olaf Willums

Massachusetts Institute of Technology, Cambridge, Massachusetts, USA

1974   Springer Basel AG

ISBN 978-3-0348-5781-9   ISBN 978-3-0348-5779-6 (eBook)
DOI 10.1007/978-3-0348-5779-6

Contents

Preface

Recent publications on the fate of our limited earth have stirred up vivid worldwide discussions, during which it became clear that our knowledge of the complex ecosystem is far from sufficient today. All planning and predictive models are only as good as the input data they are based on. Improved research methods and new ideas and impulses are therefore urgently needed, both to stimulate the specialized scientists and to supply the concerned citizen with the updated background information necessary to make responsible decisions.

Generally, most of the serious publications in the field of pollution are written by well-known and well-established scientists. It is beyond doubt that their experience and profound knowledge is an essential contribution to the understanding of our environment. But one whole group of scientists has not been able to present their ideas and concerns to a broader scientific or non-scientific readership: the young scientist at the beginning of his career. He belongs to a new generation of scientists who have studied during a time when environmental concern has become a vital part of the university life. Now, in the first decade of his academic career, his educational approach and his developing expertise in this field may synthesize ideas that will fall on what is already a fertile ground for new concepts that are urgently needed to solve the environmental crisis. Herein lies the idea of this book: a symposium by 20 scientists, all under 35, from all over the world, with educational backgrounds in a broad number of disciplines.

As the environmental problems are so complex, it would be foolish to attempt to cover the whole field with a single publication; it would be bound to be either very superficial or much too selective in the many fields to be covered. To specialize in a single field alone, on the other hand, seemed to be as unsatisfactory: the crucial problem in the environmental concern is the understanding of its interdependencies. The authors of this book have therefore tried to present their new ideas and research results of various *air pollution* aspects in depth and, most important, to stress the relations to other environmental problems. From over 50 contributions received from scientists recommended by well-known authorities in the field, 15 were selected based on the requirement both to give the specialized scientist an interesting new or somehow unconventional approach or some new research results, and at the same time to present to the interdisciplinary scientist and to the interested, concerned citizen a clear and informative introduction to the present 'state of the art' and the further research needs and prospects in various fields of air pollution. The book should therefore be of special interest to students who want both to broaden and to deepen their understanding of environmental problems.

Consequently, every article is structured in a very consistent way: the introduction defines the exact location of the field to be discussed in the often complicated web of ecological research and describes the 'state of the art' for the generally interested but nonspecialized reader; the central part of the article is an in-depth report on the individual research or on the theory developed by the author, which may, hopefully, add a creative impulse to the scientific community. The final part summarizes the research results in more general terms for the interdisciplinary reader; the author states his view of the consequences he sees in these results and describes the further research needed in his field. With this structure I hope to bring the ideas and achievements of some outstanding young scientists to the attention of a broad interdisciplinary readership without being obliged to accept the compromise of a superficial or semiscientific presentation.

I am grateful to all the well-known scientists who helped me in selecting young scientists in their fields for this publication. Without their efforts this book could not have been done. I am also greatful to Prof. HANS MISLIN of the Birkhäuser Verlag who a long time ago realized the importance of giving young scientists the chance to present their ideas to a broader readership; to Prof. ETIENNE GRANDJEAN of the Swiss Federal Institute of Technology, who sowed the first grain of environmental concern of which this book is a direct result; and to Prof. CARROLL WILSON and the World Dynamics Group at the Massachusetts Institute of Technology, who made many of us realize how strongly – and beautifully – our life, our environment and our future are interrelated.

Brussels, June 1973   JAN-OLAF WILLUMS

Foreword

WILLIAM H. MATTHEWS
Departement of Civil Engineering Massachusetts Institute of Technology

The relatively recent challenges of applying scientific knowledge to societally-perceived environmental problems pose at least three major difficulties for scientists. First, adequate response to these challenges requires the use of basic scientific theory and data in contexts that may vary significantly from those in which the original research was conducted; second, it requires imaginative and sophisticated blending of the scientific knowledge of several of the traditional disciplines; and third, it requires working in an area where societal and personal value judgements can be even more important than professional judgements in determining the definition or approach to a problem.

These difficulties face all scientists. The abilities of the 'old' and the 'young' scientists to respond are complimentary and must somehow be combined if meaningful progress is to be made. It is probably reasonable to assert that in general the older scientists are more able to overcome these difficulties on the basis of their deep and mature knowledge of their subject areas and their subtleties while their younger colleagues are more able to respond imaginatively because they are not so deeply steeped in or committed to past ways of exploring or perceiving those areas. Ideally one needs both the knowledge of the past and the willingness to venture into areas of research that seem unorthodox by the standards of the past. Thus the views of both groups must be brought to their own and each other's attention.

Any scientist encounters intrinsic resistance to his ideas in the peergroup process of certification of scientific legitimacy when his work is highly original (and thus deviates from traditional views), extends to more than one discipline (and thus may be superficial), and reflects societal values (and thus compromises scientific objectivity). The resistance to the young scientist, however, is much stronger than to his equally ventursome older colleague who has at least 'proven' himself under the age-old ground rules honored by all the scientific disciplines and subdisciplines. The young scientist therefore has a fourth difficulty to overcome – that of making his views and research ideas known.

The magnitude of this difficulty was strikingly apparent to me during the planning of the M.I.T.-sponsored interdisciplinary Study of Critical Environmental Problems (SCEP) and Study of Man's Impact on Climate (SMIC) in 1970 and 1971 and then again during the preparations for the 1972 United Nations Conference on the Human Environment. Planning groups were made up of well-established 'proven' scientists both to profit from their wisdom and to demonstrate credibility to potential sponsors and participants. When these groups established principles to guide the selection of participants and contributors there was general consensus that some young scientists should be brought into the deliberations on an equal basis with renowned experts.

There was, however, another principle which held that the outputs of these efforts should be of high scientific value and should represent the best that today's science had to offer on the subjects. Essentially this latter principle meant that the product had to have authority. Since the most effective way of demonstrating authority is to involve accepted authorities in the process, these two principles came into conflict. This, coupled with the inability to identify a very few outstanding young scientists, led to the decision in almost all cases to invite scientists who had already established their reputations. This decision had merit, but though reasonable it still may have deprived these efforts of some fresh ideas and at the very least it reinforced the system that makes it so difficult for young scientists to present their views to wide and discriminating audiences.

This volume provides a forum for some of these scientists, and such forums are sorely needed. The ideas, approaches, and results presented in these papers will be subjected to review and criticism by scientists, both old and young. Those that are sound and persuasive may be adopted and used to further science and its application to societal problems; the others may only stimulate discussion and serve to educate the author as well as the readers. No matter what its net contribution is to the advancement of science or the solution of environmental problems, this volume will have fulfilled a very important objective on the day it is published – it will have introduced some new, imaginative, and dedicated scientists to a much larger audience than they might otherwise have reached at this time in their professional development.

Introduction

As long as we do not apply the same care in testing the effects of our inventions on human life as we obviously exercise in the scientific development of these inventions, we are not mature for life in the age of technology.

CARL-FRIEDRICH von WEIZSÄCKER

Today's society seems to have a severe problem: although physically, i.e. technically adult (some may even say aging or dying), it is barely beginning to mature mentally, ethically. In VON WEIZSÄCKER'S view our society would indeed only be entering school age. We have learned as late as in the last few years to read and draw the pictures of our environmental destruction, ecologically speaking. Some have recently entered the first basic science courses and are beginning to understand the problems. Very few have so far been admitted to the final exam, where the overall understanding of the complexity of our ecosystem will be tested. The baccalaureate has not yet been given to us as a society: We may indeed not be considered a 'mature society'.

Nobel laureate DENNIS GABOR begins his newest book *The Mature Society* with an even more pessimistic description of the state of our society: 'About three-quarters of the population of the globe are still engaged in the age-old occupation of mankind, in the fight against a stingy and hostile Nature. The "most advanced" quarter has almost defeated Nature, which fights back only as a rotting corpse does, by pollution. In the rich, industrialized countries the fight has turned, almost imperceptibly, into one against human nature [1]'.

Some people might see the social unrest in the industrialized countries as obvious symptoms that the dying environment has infected the mind already. I would rather interpret it as the awakening of the mind. What it now needs is urgent assistance in order to mature and to understand the sick body before it is too late for cure. The prerequisites for mature decisions on what to do are three-fold:
1. to understand the complexity of the body,
2. to develop the remedies and
3. to forsee the long-term effects of the remedies.

The first element is what DARWIN called 'to understand the web of life'; today this has been modernized and abbreviated to the much used and often abused word 'ecology'. A pioneering scientist in this field, PAUL EHRLICH, expresses this 'understanding of the web of life' in a more emphatic way: 'What is required is no less than a revolution in human behavior, one which embodies fundamental reforms in our economic and political institutions, coupled with the wisest technological enterprises, the necessary ingredient of population control, and a new perception of man's place in nature. Since such a revolution must embrace all relationships which bind man to his fellows and to the living and nonliving environment, it is appropriate to call it "an ecolocical revolution" [2].'

The second element in the maturing process is the development of new techniques, tools and systematic approaches: the remedies for the sick body. L. SLOBODKIN has coined the very appropriate term 'ecological engineering' [3]. Ecology has to develop as a *whole* science: The various branches must get together in order to develop new solutions, new systems. Not only technical tools for 'repairing' the environment should be understood under the term 'ecological engineering' but also accurate methods for monitoring it and controling the sources. And as the understanding of the 'pure' sciences of physics and chemistry has given birth to modern engineering, the understanding of the 'pure ecology' will enable the 'eco-engineer' to develop the practical means to solve the eco-crisis.

The third element, then, is the understanding of the long-term behavior of the system. A primary sign of maturity is the ability to plan ahead and to foresee difficulties and obstacles before they become crucial. Only recently some serious efforts to develop a system necessary for evaluating the usefulness of environmental policies have been undertaken. JAY FORRESTER's pioneering work in system dynamics which enables the analysis of complex feed-back systems has shown the direction in which further research has to go. Its application to a whole world system, 'world dynamics', has evoked ardent discussions. I personally believe that the prime objective of the authors of *The Limits to Growth*, where the world model was presented to a broad readership [4], was just to evoke this discussion: to force the society to get involved in looking into the future, to expand the time horizon of planning, or at least to include a thought on the 'possible later consequences' in today's decisions. It is evident that these planning tools are far from perfect; there is still a long way to go. But I am somehow confident. Didn't the worldwide discussion of both the American *The Limits To Growth* and the British *A Blueprint for Survival* [5] somehow prove that mankind is maturing, and is rejecting its ignorance of the future? More and more of us begin to understand and 'feel' the truths in BUCKMINSTER FULLER's definition of our living environment: The limited Spaceship Earth.

Two crucial problems have become clearer in the vivid discussions of the 'futurologists': one is the lack of a coherent long-term goal for our society. Few people have taken time to think about where we really should direct our future to, if we can influence it (what all planners, by definition must believe). Both government and business plan only for relatively short time spans, seldom exceeding a decade or two. But planning in a system with dynamic fluctuations of the order of several decades requires at least a general time horizon of one or two generations. Where do we really want to go? Books like DENNIS GABOR's *The Mature Society* or GARRET HARDIN's *Exploring New Ethics for Survival* [6] have opened up this vital question for discussion: a discussion which is of prime importance. A society without a sound goal is bound to become decadent. And a growth of the GNP alone is obviously not a sound goal. DENNIS GABOR defines his mature society as a peaceful world on a high level of material civilization, which has given up growth in numbers and in material consumption, but not growth in the quality of life, and one which is compatible with the nature of homo sapiens.

The second problem is the impossiblity of planning for any goal with an insufficient knowledge of the present status. The conclusion of every planning model, as sophisticated as it may be, is only as good as the information supplied as an input. And so we return again to the first element in our analysis: the urgent need for an improved understanding of our present ecosystem. Both the knowledge in 'pure ecology' and applied 'eco-engineering' has to be increased drastically, if we want to be able to plan our future, at least to some extent. We need urgently interdisciplinary contributions in order to progress in all three fields. Herein lies the basic idea of this book. I believe that there is a great potential of new ideas and fresh approaches among the world's younger scientists on how to understand and improve our environment. These ideas must be transmitted both to the concerned citizen and to the specialized scientist. We all belong to the same 'spaceship-society', bound, hopefully, for maturity.

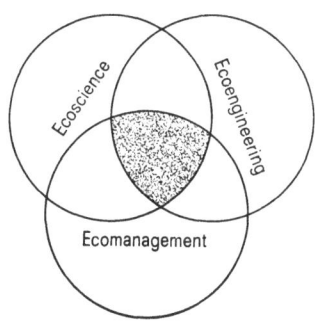

References

[1]   D. GABOR, *The Mature Society* (Praeger Publishers, New York 1972).
[2]   P. EHRLICH, J. P. HOLDREN, *Global Ecology; Readings Toward a Rational Strategy for Man* (Harcourt, Brace, Jovanovich, Inc., New York 1971).
[3]   L. SLOBODKIN, *Aspect of the Future in Ecology*, in: *The Ecological Conscience: Values for Survival* (Prentice Hall, 1970).
[4]   D. MEADOWS et al., *The Limits To Growth* (Universe Books, New York 1972).
[5]   E. D. GOLDSMITH et al., *A Blueprint for Survival*, in: *The Ecologist*, Vol. 2, No. 1 (London, January 1972).
[6]   G. HARDIN, *Exploring New Ethics for Survival: The Voyage of the Spaceship Beagle* (Viking Press, New York 1972).

Ecoscience I:
The Pollution – Biosphere Interface

In the past man was able to regard the air as a limitless and undestructible resource, and in economical terms, as a 'common'. Only recently did we begin to realize how fragile our environment is: the atmosphere is in a very delicate balance, which we eventually might upset with our activities. Even a change of a few percent in global albedo may be sufficient to cause extensive changes in the fields of polar ice. STEPHEN H. SCHNEIDER of the National Center for Atmospheric Research in Boulder, Colorado (USA), investigates in his contribution the equilibrium temperature of the earth and the impact of man's activities. It has long been suspected that suspended atmospheric particles might affect this balance. By contrasting certain models, he illustrates some major problems in modeling studies and suggests that the development of refined methods to analyze the behavior of the complex climatic feedback mechanisms of the land-ocean-atmosphere system is most urgently required.

Acute effects of air pollution are usually experienced close to the source of the emission. As a consequence, most research work and control measures have dealt, in the past, and continue to deal with air pollution in areas close to the source. However, an awareness is rapidly being gained of the importance of the transport of pollutants over large distances and the subsequent removal into previously unpolluted forests, soils and oceans. Direct consequences are, for example, increased global levels of CO_2, acidification of lakes and soils and accumulation of pesticides in the marine environment. The current theories on moderate- and long-range transport of air pollutants and their removal from the atmosphere are discussed by D. M. WHELPDALE and R. W. SHAW. Their latest measurements of the sulphur oxide transport over Lake Ontario stresses not only the need for better control on an international level, but poses a number of additional questions that PETER H. SAND answers in his article on trans-frontier pollution and international law.

How seriously the question of long-range transport of pollutants must be taken becomes evident when reading the biologist SATU HUTTUNEN's contributions on air pollution effects in northern Finland. He has found indications of serious danger in regions of 'untouched natural beauty', the still magnificent woods of northern Finland. This unplanned and often unnoticed destruction of forests all over the world is extremely regrettable because forests are the greatest achievement of ecological evolution. They are the largest, most complex, and most self-perpetuating of all ecosystems. It is in forestry that man has the best opportunity to work *with* nature, not against nature. As biological systems react very early to environmental pollution, the use of biological parameters will become a very important supplementary technique to the physical and chemical methods of environmental monitoring; SATU HUTTUNEN proposes a system based on comparative microanalysis of pine needles.

How does a high concentration of pollutants affect our own biological system? Many medical papers have been written on this subject; most of them are either based on statistical investigations or simulated laboratory tests. A Yugoslav team under PAVLE TODOROVIĆ, has chosen another route and investigated the air pollution problem where the consequences may be fatal: inside an automobile driving through the heavy traffic of Belgrade, Yugoslavia. He concludes that the psychophysical capability of a driver is significantly reduced by the concentrations of CO and NO_x he measured inside the car and at busy cross-roads, and the thus air pollution may have a far greater impact on traffic accidents than has been generally assumed.

It is often believed that certain pollution aspects, like photochemical smog, may only be expected at some few places in the world with exceptional climatologic features, like Los Angeles or Mexico City. But a Portuguese research project by S. J. FORMOSINHO and A. C. CARDOSO came to the conclusion that the symptoms in Europe indicate an increase in the kind of photochemical smog that already mildly exists in several European cities. They discuss today's theories on photochemical smog formation and present a new theory to explain the rapid oxidation of NO to NO_2 in that process. The paper is another example showing the importance of basic research; without this progress in ecoscience the new technical means developed by eco-engineering may never become fully effective. The only way to cope with the problem is the inter-disciplinary approach.

Atmospheric Particles and Climate:
The Impact of Man's Activities

by Stephen H. Schneider
National Center for Atmospheric Research, Boulder, Colorado, USA[1]

Abstract

The equilibrium temperature of the earth is maintained by a balance between the unreflected part of the incoming solar energy, which is absorbed by the earth-atmosphere system, and the outgoing longwave radiation escaping from the earth to space. It has long been suspected that suspended atmospheric particles (aerosols) might affect this balance, primarily by affecting the albedo or reflectivity of the earth, thereby altering the amount of solar energy absorbed by the earth. In light of some recent evidence suggesting the existence of an increase in atmospheric particle concentrations (presumably related to man's activities), the need for development of adequate numerical models to study this problem is apparent. Recent numerical models studying the effect of particles on climate are often based on multiple scattering radiative transfer calculations, and use global averages for particle concentrations and optical properties. By contrasting certain existing models, some major problems in modeling studies that attempt to answer the question of the effects of increased atmospheric particles on climate can be illustrated. It will also be apparent that another uncertainty in the results of such studies arises from a lack of adequate observed input data on the geographic and vertical distributions of particle concentrations and their optical properties. Furthermore, a model that could realistically simulate the impact of increasing atmospheric particle concentration on climate must eventually include the simultaneous coupled effects of all the important atmospheric processes, such as fluid motions and cloud microphysics, in addition to the radiative transfer effects. Current modeling studies already do predict that increases in particle concentrations could have a significant effect on climate. Now, it remains for us to develop the kinds of refined models needed to verify or deny these predictions.

1   Introduction

It is well established that the climate of the planet earth has undergone significant changes in the past, and there is every reason to believe that there

---

[1] The National Center for Atmospheric Research is sponsored by the National Science Foundation.

will be other changes in the future. In the last century it is possible to document
an increase of about 0.6 °C in the mean global temperature between 1880 and
1940 and a subsequent fall of temperature by about 0.3 °C since 1940 (MIT-
CHELL [1]). In the polar regions north of 70° latitude the decrease in temperature
in the past decade alone has been about 1 °C, several times larger than the global
average decrease (see Fig. 3.8 in the SMIC Report [2]). Up till now, past climatic
changes (except possibly those of the last few decades) could hardly have been
caused by man's activities. However, we have recently realized that man has
altered the face of the earth and the composition of the atmosphere on such a
large scale that his influence can no longer be ignored relative to nature's.
Numerous studies have been devoted to various aspects of the question of the
possibility of inadvertent modification to the climate by man's activities.
Among the most recent are the SMIC Report [2], the SCEP Report [3],
SCHNEIDER and KELLOGG [4] and SINGER [5].

Two of the most publicized means by which it is thought man could
influence climate are related to the dumping of CO_2 and 'dust' particles (or
gases such as SO_2 which are subsequently converted to atmospheric particles)
into the atmosphere, from activities that are usually associated with the burning
of fossil fuels. The effect of increased CO_2 is well-known to be an increase in the
temperature of the earth's surface by an enhancing of the 'greenhouse effect'
(MANABE and WETHERALD [6]). Particles, on the other hand, could decrease the
temperature of the earth's surface by screening out part of the incoming solar
beam and thus raise the albedo (or reflectivity) of the earth – thereby decreasing
the amount of solar enerby absorbed by the earth (McCORMICK and LUDWIG [7];
RASOOL and SCHNEIDER [8]; YAMAMOTO and TANAKA [9]). Some scientists
already feel that particles might be responsible for the recently observed decrease
in the earth's surface temperature, and that increased particulate pollution

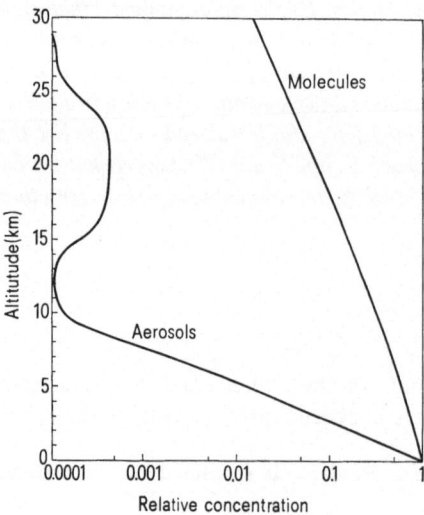

Figure 1
Vertical distribution of the relative
concentration of particles and air
molecules (ELTERMAN [14]).

could lead to a global ice age (BRYSON [10]). However, others have suggested that the effect of increased amounts of suspended atmospheric particles (also called atmospheric aerosols) could as well be a *warming* of the climate (CHARLSON and PILAT [11]; MITCHELL [12]); and GUCCIONE [13] has even questioned whether there is, in fact, a build-up of particles in the atmosphere. This paper will attempt to summarize the 'state of the art' concerning the question of atmospheric aerosols and climate, and will try to outline what additional kinds of observational and theoretical experiments remain to be performed before we can adequately determine the effects of particles on global climate. Emphasis in this article will be placed on tropospheric particles, since the sources for these are primarily near the earth's surface. In fact, the vast bulk of all suspended atmospheric particles reside in the troposphere (up to about 10 km), as can be seen in Figure 1 (ELTERMAN [14]). The source of the stratospheric particle load, though still somewhat of a controversy, might well be the natural injection of sulfur dioxide gas into the stratosphere by volcanic eruptions (see the chapter by FRIEND in RASOOL [15]). This paper will draw heavily on the chapter *The Chemical Basis for Climate Change* by S. H. SCHNEIDER and W. W. KELLOGG in the book *Chemistry of the Lower Atmosphere*, edited by S. I. RASOOL [15]. This paper also appears in a modified form in the 1972 Proceedings of the 18th Annual Technical Meeting of the Institute of Environmental Sciences and also has been published in Quaternary Research, Vol. 2, 1972.

2   Trends in Particle Loading of the Atmosphere

The geographic distribution of the concentration of particles in the atmosphere is determined by a variety of sources of particle production and sinks of particulate removal. A major source of atmospheric particles (the data described in this section is from Table 8.1 of the SMIC Report [2] and is presented graphically in Fig. 3 of MITCHELL [1]) is soil and rock debris picked up and carried by the wind. Although the latter is a 'natural' source of particles, this mechanism is clearly enhanced by deforestation and agricultural practices of man. Other major contributors are sea salt (from sea spray droplets that evaporate), from forest fires and other vegetation fires, from periodic volcanic eruptions, and by conversion of certain gases to particles in the atmosphere. Two such substances which are produced both naturally and by man's activities are sulfur compounds and hydrocarbons. These start out as gases, but can subsequently be converted into atmospheric particles. The 'blue haze' often observed in the Smoky Mountains of Tennessee is due, in large measure, to the exudations of hydrocarbons by vegetation (WENT [16]). Most of the 'man-made' particles that become part of the total atmospheric particle loading are initially in the form of sulfur dioxide gas that is released into the atmosphere.

The man-made contribution is from 5 to 45 percent of the total *mass* of suspended particles. The spread in this ratio is an indication of the lack of precise data available to us on particle sources. But, in midlatitudes of the northern hemisphere, where most of the human activity that produces these emissions takes place, it is probable that in this large belt the man-made particle production is comparable to nature's.

In addition to the increase in the mass of the total particulate load, there is another important consequence of the injection of man-made particles into the atmosphere. This stems from the fact that the average size of man-made particles is smaller than the average size of naturally produced particles. That is, when the man-made contribution is included with the natural or 'background' distribution of particles, the average size of the suspended particles is decreased (see Fig. 8.2 of the SMIC Report [2]). This is important because it indicates that the man-made contribution can increase the total *number* density of atmospheric particules by a greater factor than the corresponding increase in particle *mass* density (since mass density depends upon the average radius cubed).

The sinks for the removal of particles can be classified (in inverse order of importance) as a) dry sedimentation or fallout, b) impaction against the ground, c) rainout (snowout) – where particles are incorporated into cloud droplets in the cloud, and d) washout – in which falling rain or snow scavenges the particles below cloud level. Clearly, the smaller particles will travel further and remain longer in the atmosphere than the larger ones. The average tropospheric residence time of the small, but highly abundant, particles (with radii of about 0.1 μm) is of the order of one month – enough time for many of these to make a trip around the world. These removal processes, which are highly variable and depend upon local meteorological conditions, are discussed in more detail in the chapters by CADLE, FRIEND, HIDY and PRUPPACHER in RASOOL [15].

The intensity and location of particle sources and sinks determine the geographic distribution of particles around the globe. Although it is apparent that our knowledge of the sources and sinks of particles, and consequently the geographic distribution of particulate loading, is still very unsatisfactory, enough studies have been made to establish a tentative estimate of particle concentration trends on a hemispheric basis. Based on world-wide data of the increase in atmospheric turbidity (i.e., attenuation of the solar beam due to scattering and absorption by particles) and shipboard measurements of electrical conductivity far out in the Atlantic Ocean, C. JUNGE and the SMIC working group on particles were able to estimate for the northern hemisphere that 'since 1910 the total number concentration in much of the northern hemisphere has *increased* by a factor of about 2, that the average size has *decreased* by a factor of about 1.5, and that the mass concentration has perhaps *increased* by a factor of about 1.5' (p. 205 of the SMIC Report [2]). No indications of global pollution by particles in the southern hemisphere is yet apparent. This increase in atmospheric particle loading in the northern hemisphere has apparently taken place in spite of efforts to reduce smoke emission in the cities (LUDWIG, MORGAN and MCMULLEN [17]).

In connection with the last point, the reduction in smoke from large coal-burning power plants and steel mills has been achieved by precipitating

the larger particles that are the most obvious to the eye, but the smaller particles of submicron size still largely escape. The larger ones would have been
removed first from the atmosphere, so the escaping particles are just those
that last longer and contribute most to the hemispheric turbidity trend wherever
they are. This is borne out by Figure 2, where the urban mass concentration of
particles in the U.S. has gone down as a result of air pollution control efforts,
while it has gone up at the nonurban stations. In addition, the continued
release of SO_2 into the atmosphere, which is expected to increase globally (see
KELLOGG et al. [18] and Fig. 15 of LUDWIG, MORGAN and McMULLEN [17]) over
the next several decades, is a prolific source of particles. Since global climatic
effects are dependent on the *global* energy balance, which is essentially an
'area-weighted' phenomenon, trends in the rural and oceanic particle loadings
are most significant for climatic influences.

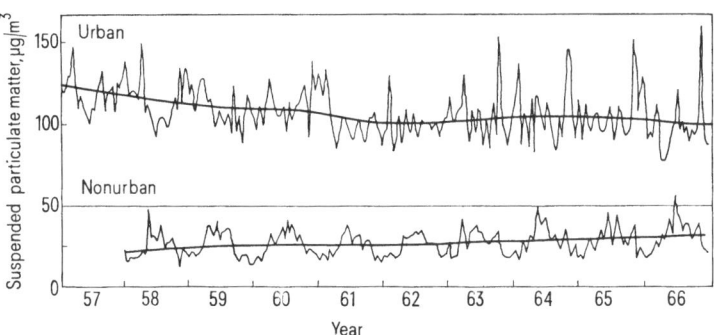

Figure 2
Long-term trends at 58 urban and 20 nonurban sites. The jagged curves are the
averages for each sampling interval; the smoother trend curves were calculated using a
technique that combined weighted measurements of curve fit and smoothness. The
average increase at the nonurban sites during this period is a steady 3 to 4 percent
(LUDWIG, MORGAN and McMULLEN [17]).

3   Particles and the Radiation Balance

The radiative equilibrium temperature of the earth is maintained by a
balance between the unreflected part of the incoming solar energy that is
absorbed in the earth-atmosphere system and the outgoing terrestrial infrared
flux emitted to space by the earth's surface, optically active gases, clouds, and
atmospheric particles. If, for example, the albedo of the earth is increased more
sunlight will be reflected to space, and the planetary radiative equilibrium
temperature will be decreased. The average planetary *radiative equilibrium*
temperature is about —20 °C, whereas we know the *surface* temperature is
about +14 °C. The difference between the two values is due to the presence
of our atmosphere, and is related to the well-known phenomenon called the
'greenhouse effect'.

The average surface temperature is higher than the effective radiative temperature because the atmosphere is semitransparent for solar radiation but nearly opaque to infrared radiation as a result of infrared absorbing gases and clouds. Thus, the surface, which absorbs much of the solar radiation, becomes a heat source for the lower atmosphere, which on the average cools steadily with altitude to about 10 km. This part of the atmosphere is called the 'troposphere'. The *observed global average* tropospheric vertical temperature 'lapse rate', $\partial T/\partial Z = -6.5\,°K/km$, is determined by both radiative heating and convection processes.

Since the observed global-average lapse rate is believed to remain nearly constant in the troposphere (MANABE and WETHERALD [6]) it is usually concluded that a change in the mean planetary radiative temperature will also result in a similar variation (though of different magnitude) in the surface temperature.

Thus a one percent increase in global albedo would result in about a 2°C decrease in surface temperature (BUDYKO [19]), should other factors remain unchanged.

The effect of an aerosol layer located just above the earth's surface on the albedo is shown in Figure 3 (from SCHNEIDER [20]). Here, α_{BS} is the fraction of solar flux backscattered to space from the aerosol layer, α is the fraction absorbed, and α_S is the albedo of the underlying surface. The transmission of radiation through the particle layer is $t = 1 - \alpha_{BS} - \alpha$. Then α_S of the transmitted beam is reflected from the surface (neglecting directional effects of the particles in changing α_S), with t times this amount escaping to space through the aerosol layer. Thus, the total or effective albedo, α_E, of the combined

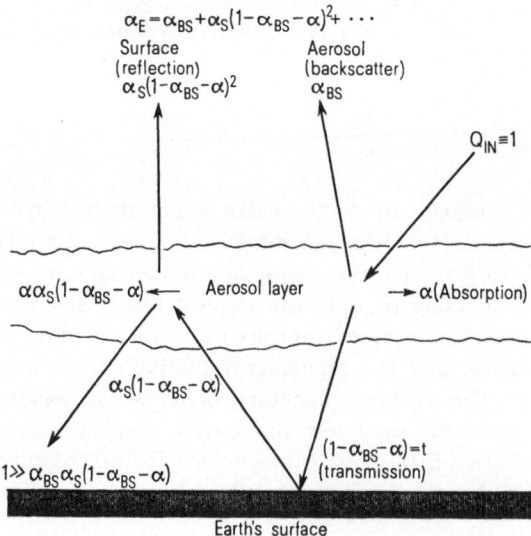

Figure 3
Schematic diagram showing how the presence of an absorbing and scattering aerosol layer in the lower atmosphere can change the effective albedo of the earth's surface (SCHNEIDER [20]).

aerosol-surface system is obtained by the summation of the multiple reflections:

$$\alpha_E = \alpha_{BS} + \frac{t^2 \alpha_S}{1 - \alpha_S \alpha_{BS}} \tag{1}$$

Using (1) we find that when $\alpha_{BS} = \alpha = 0.05$, the effect of this aerosol is to raise the surface albedo if the surface albedo were initially less than about 0.28. Since the albedo of the cloudless fraction (which comprises nearly 50 percent) of the surface area of the earth is about 0.14 on a global average (LONDON and SASAMORI [21]), the global effect of an aerosol layer with absorption comparable to backscatter would appear to be an *increase* in the albedo of the cloudless fraction of the earth. This question is discussed in more detail below.

In order to determine α and α_{BS} for an aerosol layer multiple scattering radiative transfer calculations can be made, using as input data the size distribution of the particles and the particle optical properties (i.e., the real part of the index of refraction, n_r, and the index of absorption, n_i, or imaginary part of the index of refraction). These quantities depend strongly upon the shape and chemical composition of the aerosol particles, and their values should be determined by *in situ* atmospheric measurements whenever this procedure is possible. (This is desirable since the optical properties of an aerosol depend also on the local atmospheric relative humidity). The size distribution for particles with radii between 0.1 μm and 10 μm has been determined by JUNGE [22] to be such that relative number density of particles with given radii is roughly proportional to the inverse fourth power of the particle radius. Thus, there is a factor of 10^8 more particles with radii near 0.1 μm than with radii near 10 μm. This is due to the previously mentioned fact that larger particles are removed more easily from the total particle population than the smaller ones.

Multiple scattering calculations have been performed to determine quantitatively the parameters α_{BS} and α in the papers of RASOOL and SCHNEIDER [8] and YAMAMOTO and TANAKA [9] (as described in chap. 8 of SMIC [2]). In these analyses, it was necessary to assume values for the optical properties of the aerosols, the size distribution of the particles, and the effect of the particles on the part of the earth covered by clouds. In both studies, the JUNGE size distribution was used and the particles were assumed to have no effect on either the amount of cover, height, or albedo of the cloudy half of the earth – since most of the tropospheric particles are found in the first few kilometers of the troposphere (see Fig. 1). In these independent studies, a real index of refraction, $n_r = 1.5$, was assumed, and two values of the imaginary part of the index of refraction or index of absorption, n_i, were given, $n_i \approx 0.00$ and $n_i \approx 0.01$; the latter implying that the fraction of radiation flux absorbed by the aerosol is comparable to the fraction that is backscattered for a median visible wavelength of 0.55 μm. For the infrared, a median wavelength of 10 μm and an absorption index of $n_i \approx 0.1$ was used by RASOOL and SCHNEIDER, whereas YAMAMOTO and

TANAKA neglected infrared effects. In the results of RASOOL and SCHNEIDER given in Figure 4, the infrared flux to space in the presence of the aerosol layer is computed, but the effect of an increase in aerosol optical thickness on reducing the infrared radiation ($\lambda = 10\,\mu m$) to space is found to be relatively less significant than the resulting increase in global albedo (for $\lambda = 0.55\,\mu m$), for both cases of n_i. This is a consequence of the size distribution as mentioned previously and also is a consequence of the fact that most particles are in the first few kilometers of the troposphere (see Fig. 1). The global average albedo of the cloudless fraction of the earth for present conditions ($\tau_{VIS} = 0.1$) was taken to be 0.10. By increasing the visible optical thickness of the particle layer, τ_{VIS}, the effective albedo of the earth's surface, α_E, was found to be increased. The computed increase in the albedo of the cloudless part of the earth results in a decrease in the amount of solar flux absorbed by the earth-atmosphere system. The intersection of the absorbed solar flux curves with outgoing infrared flux values determines the global average equilibrium surface temperature as a function of aerosol optical thickness, which is cross-plotted on Figure 4b. The single scattering albedo, $\bar{\omega}_0$, which depends upon n_i, is given in the figure for the two cases of index of absorption assumed in the calculations, $n_i \approx 0.00$ ($\bar{\omega}_0 = 0.99$) and $n_i \approx 0.01$ ($\bar{\omega}_0 = 0.90$).

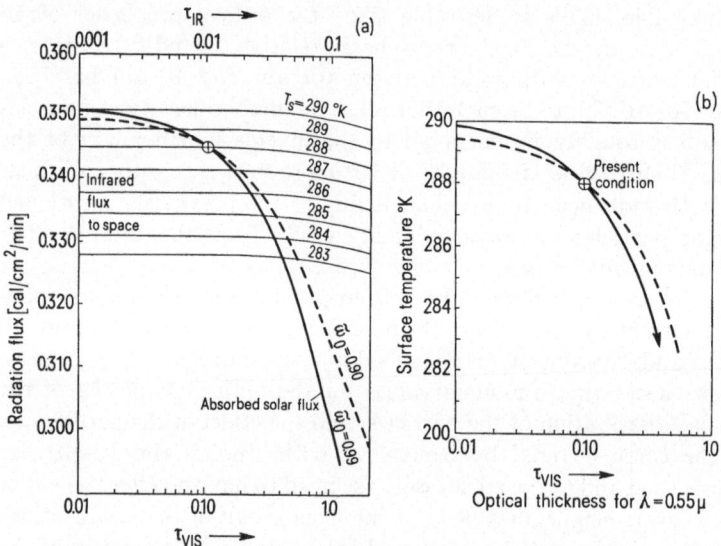

Figure 4

In (a), both the global average values of absorbed solar radiation flux and terrestrial infrared radiation flux to space are plotted as a function of increasing accumulation of aerosols in the atmosphere. The absorbed solar flux decreases with increasing optical depth for both assumed values of the single scattering albedo parameter, $\bar{\omega}_0$. The infrared flux to space, calculated for several values of surface temperature, is practically unaffected by increasing aerosols, as explained in the text. The intersections of infrared and visible flux values determine the equilibrium global surface temperature for a given optical thickness, τ_{VIS}, and are cross-plotted in (b) (RASOOL and SCHNEIDER [8]).

The computations show that an increase in aerosols by a factor of 4 could raise the albedo of the cloudless part of the earth sufficiently to decrease the surface temperature by 3.5 °C (assuming fixed tropospheric lapse rate).

The computations of YAMAMOTO and TANAKA agree with those of RASOOL and SCHNEIDER in predicting a significant cooling of the earth from an increase in aerosols. However, these results depend, of course, on the major assumptions of the models: global average (i.e., homogeneous) horizontal distribution of model variables; assumed values for aerosol optical properties (particularly n_i); and finally, the non-interaction of the particles with cloudiness and the atmospheric circulation. This latter point will be the subject of the next section.

As for the question of optical properties, equation (1) shows that the cooling/warming effect of an aerosol layer depends upon the backscatter/absorption ratio of the particles for sunlight for a given surface albedo. ENSOR et al. [23] using a single scattering-model show that this ratio is highly sensitive to the index of absorption, n_i. This author has recently repeated the RASOOL and SCHNEIDER computations and found that warming replaces cooling in their model only when n_i gets larger than about 0.1. Unfortunately, n_i is a very difficult quantity to measure *in situ* in the atmosphere, and estimates range all the way from $n_i \sim 0$ for 'clean' water droplets (which have almost no solar absorption) to $n_i \sim 0.1$ (which would lead to significant warming). Large soot particles found near heavily industrialized areas tend to have a larger value of n_i than naturally occurring particles (EIDEN [24]; FISCHER [25]). On the other hand, recent measurements in Los Angeles smog show $n_i \sim 0.005$ (SHETTLE and WEINMAN [26]).

From this discussion we can see the possibility that an increase in total particle loading arising from man-made particles could also change the prevailing value of n_i, but any such contention requires a *continuous* observational record of particle concentrations and optical properties taken at *many* different places on earth in order to be proven.

Another question remaining to be better resolved is that of the vertical distribution of particles – since an aerosol layer located above the cloud tops (which have albedos much larger than most underlying surfaces) could lower the effective albedo of the cloud tops (see the comment by CHARLSON, HARRISON and WITT [27] and the reply by RASOOL and SCHNEIDER [28]). It is not clear how the addition of man-made particles will affect the shape of the vertical distribution (Fig. 1). Finally, a major difficulty inherent in all global-average models of the radiation balance is that they do not consider possible coupled interactions between changes to the radiation balance from increased aerosols, and a consequent dynamic response by the atmosphere or oceans. Determination of this coupling requires the use of large-scale numerical models, as discussed in the next section.

Nevertheless, these global-average models do indicate that changes in atmospheric aerosols can significantly affect climate. This suggests that programs for the monitoring of the geographic distribution of particle concentrations and optical properties be implemented, along with the improvement of mathematical models, incorporating as many coupled effects or 'feedback mechanisms' as can be included with the computing machinery that will be available to us.

4   More General Climate Models

In order to study the sensitivity of the climate to changes in the concentration of a certain pollutant, it is first necessary to determine the initial effect of increases in the pollutant on the radiation balance. After this, it must be established how other coupled or interacting processes (often called feedback mechanisms) might act to dampen (negative feedback) or amplify (positive feedback) that initial effect on the climate.

The radiation balance is strongly coupled with the atmospheric motions. The rates of solar heating and infrared cooling are highly variable both over the globe and throughout the vertical extent of the atmosphere. This unequal or differential heating of the globe, when coupled with the rotation of the earth, is the driving force behind the motions we recognize as winds and ocean currents. These motions, both horizontal and vertical, regulate the distribution of temperature, cloudiness (which also requires suitable particles acting as cloud nuclei) and precipitation over the globe.

The motions of the system result in a transport of heat from areas of positive radiation balance (the equatorial regions) to areas of negative radiation balance (the polar regions). In this process the jet streams, trade winds, eastward winds of the midlatitudes, westward winds of the polar latitudes, and migratory large-scale weather systems or eddies are generated in the atmosphere (LORENZ [29]).

The studies of MITCHELL [12] concern one such example of the importance of understanding the possible coupled effects between changes in the atmospheric radiative heating rates due to particles in the atmosphere and other heat balance variables. MITCHELL has suggested that, by including the effect of the evaporability of the earth's surface underlying an aerosol layer, a *warming* of the surface layer is possible, even if the aerosols *increase* the planetary albedo.Thus, it is implied that even if the optical properties of aerosols are reliably known, and we could therefore determine their effect on the planetary albedo, we would still be unable to predict the effect of aerosols on *surface* temperature by merely assuming an inverse relationship between changes in the planetary albedo and the response of the surface temperature, as has been assumed in the past by most other authors.

This conclusion is reached because MITCHELL assumes that an increase in the radiative heating rate in the lower atmosphere (that might be expected from the increase in the absorption of sunlight by a low-lying aerosol layer) could increase the tropospheric temperature (and lapse rate) near the surface However, SCHNEIDER [30] has argued that such an increase in the surface temperature might not be realized on a large scale because of the apparent tendency of the atmosphere to conserve the average tropospheric lapse rate.

Thus, when attempting to consider coupled effects, it is obviously necessary to include in a model all the important mechanisms that could significantly affect the result. Such a task eventually requires the use of large-scale numerical models of the land-ocean-atmosphere system, including as many details of fluid dynamic and cloud forming processes as the speed and size of the available computing machinery will allow. Meanwhile, we must study these coupled effects one at a time.

Another important and widely discussed example of a coupled effect or climatic feedback mechanism is the snow-and-ice-cover-albedo-temperature coupling, where the high reflectivity of snow and ice as compared to water and land surfaces plays a dominant role in the climate of polar regions. However, the extent of the snow and ice cover of the earth's surface strongly depends upon the surface temperature. Thus, a lowering of the planetary temperature would permit increased snow and ice cover, which would increase the planetary albedo – causing a further decrease in solar energy absorbed in the earth-atmosphere system – thereby further lowering the temperature. This positive feedback mechanism has caused many scientists (e.g., BUDYKO [31]; SELLERS [32]) to construct models that consider the question of the permanency or stability of the polar ice caps, since changes from the present extent of ice coverage would appear to 'feed' on themselves.

Numerical experiments with semiempirical models of the heat balance of the earth-atmosphere system (including the effect of temperature-albedo coupling) suggest that changes in the planetary temperature of greater than a few degrees or variations in energy input to the earth larger than a few percent would appear, if sustained over a sufficiently long period, to be sufficient to result in either a rapid expansion or complete disappearance of the polar ice sheets (see the discussion of semiempirical climate modes in SCHNEIDER and KELLOGG [4], and in chap. 6 of the SMIC Report [2]). However, this ice-and-snow-cover-albedo-temperature coupling positive feedback mechanism discussed here must also be viewed in the light of still other coupled processes that might modify these conclusions. (See also the discussion in SCHNEIDER and GAL-CHEN [33]). For example, hydrological processes should also be included in the feedback loop, since ice and snow cover are merely the solid phase of water. Thus, the strong positive link between the extent of snow and ice cover and the local temperature assumed in the semiempirical models discussed above will be effective only insofar as there is an appropriate amount of precipitation – to build up continental glaciers, for example (DONN and EWING [34]. This case is a good example of the general complexity of the land-ocean-atmosphere system, since the hydrological processes depend in part upon the physical and chemical condition of the earth's surface, the dynamic state of the oceans and the atmosphere, and the global cloudiness (which also has a significant effect on global albedo). The possible role of cloudiness as a climatic feedback mechanism has been summarized recently by SCHNEIDER [35].

5   Conclusion

The preceding discussion could be summarized by the following main points:

1. An increase in the number concentration and mass of suspended atmospheric particles as well as an upward trend in turbidity and a decrease in the average particle size can be documented for much of the area of the northern hemisphere. These trends are almost certainly a result of man's activities in the northern hemisphere. However, detailed knowledge of the intensity and location of sources and sinks of particulate matter as well as the geographic and vertical distribution of suspended atmospheric particles is still very unsatisfactory.

2. Global-average models of the radiation balance using reasonable estimates for aerosol optical properties indicate that an increase in the particle loading of the atmosphere could significantly increase the albedo of the earth – presumably resulting in a decrease in the temperature of the earth's surface. Even a change of a few percent in global albedo may be sufficient to cause extensive changes in the fields of polar ice. However, better data for aerosol optical properties must be obtained. Furthermore, there is a possibility that the addition of man-made particles to the existing particle background could also influence the optical properties of all the suspended atmospheric particles combined. This suggests that monitoring programs be implemented at many different locations on the globe in order to obtain long-time trends in both the concentration and optical properties of aerosols.

3. Large-scale numerical models of the land-ocean-atmosphere system should be further developed to determine how changes in the radiation balance from increased aerosols might evoke simultaneous variations in other atmospheric variables. Only with such detailed models can we predict what the ultimate effect of increases in atmospheric particles will be on the global climate.

Despite the simplicity inherent in the modeling studies available to us at present, we should not dismiss their predictions glibly. Rather, it remains for us to proceed rapidly with the development of the kinds of refined models and measurements needed to verify or deny these predictions.

Acknowledgment

This work was initially supported by the National Academy of Sciences/National Research Council through a Resident Research Associateship, at the Institute for Space Studies, Goddard Space Flight Center, NASA, New York, New York, 10025, and was completed when the author was a GARP Fellow in the Advanced Study Program of the National Center for Atmospheric Research.

References

[1] J. M. MITCHELL, Jr., *Quaternary Research 2*, 436 (1972).

[2] Study of Man's Impact on Climate (SMIC), *Inadvertent Climate Modification* (M.I.T. Press, Cambridge, Mass. 1971).

[3] Study of Critical Environmental Problems (SCEP), *Man's Impact on the Global Environment* (M.I.T. Press, Cambridge, Mass. 1970).

[4] S. H. SCHNEIDER and W. W. KELLOGG, *The Chemical Basis for Climate Change*, in S. I. RASOOL: *Chemistry of the Lower Atmosphere*, chap. 5 (Plenum Press, New York, New York 1973.

[5] S. F. SINGER, Ed., *Global Effects of Environmental Pollution* (Reidel Publishing Co., Dordrecht, Holland; Springer-Verlag, New York 1970).

[6] S. MANABE and R. T. WETHERALD, J. atmos. Sci. *24*, 241 (1967).

[7] R. A. McCORMICK and J. H. LUDWIG, Science *156*, 1358 (1967).

[8] S. I. RASOOL and S. H. SCHNEIDER, Science *173*, 138 (1971).

[9] G. YAMAMOTO and T. TANAKA, J. atmos. Sci. *24*, 1405 (1972); also see chap. 8 of the SMIC Report [2]).

[10] R. A. BRYSON, Weatherwise *21*, 56 (1968).

[11] R. J. CHARLSON and M. J. PILAT, J. appl. Met. *8*, 1001 (1969).

[12] J. M. MITCHELL, Jr., J. appl. Met. *10*, 703 (1971).

[13] E. GUCCIONE, *Is Pollution a Problem? No, Breathe Easier*, New York Times, p. 25 (28 August 1971).

[14] L. ELTERMAN, Appl. Opt. *3*, 745 (1964).

[15] S. I. RASOOL, Ed., *Chemistry of the Lower Atmosphere* (Plenum Press, New York, New York 1973.

[16] F. W. WENT, Proc. nat. Acad. Sci. *46*, 212 (1960).

[17] J.H. LUDWIG, G. B. MORGAN and T. B. McMULLEN, *EOS*, Trans. Am. geophys. Un. *51*, 468 (1970), reprinted as chap. 25 in: W. H. MATTHEWS, W. W. KELLOGG and G. D. ROBINSON, *Man's Impact on the Climate* (M.I.T. Press, Cambridge, Mass. 1971).

[18] W. W. KELLOGG, R. D. CADLE, E. R. ALLEN, A. L. LAZRUS and E. A. MARTELL, Science *175*, 587 (1972).

[19] M. I. BUDYKO, Tellus *21*, 611 (1969).

[20] S. H. SCHNEIDER, J. appl. Met. *10*, 840 (1971).

[21] J. LONDON and T. SASAMORI, *Radiative Energy Budget of the Atmosphere*, in: W. H. MATTHEWS, W. W. KELLOGG and G. D. ROBINSON, *Man's Impact on the Climate*, (M.I.T. Press, Cambridge, Mass. 1971); also Space Res. *XI*, 639 (1971).

[22] C. JUNGE, *Atmospheric Chemistry and Radioactivity* (Academic Press, New York 1963).

[23] D. S. ENSOR, W. M. PORCH, M. J. PILAT and R. J. CHARLSON, J. appl. Met. *10*, 1303 (1971).

[24] R. EIDEN, Appl. Opt. *10*, 749 (1971).

[25] K. FISCHER, Beitr. Phys. Atmos. *43*, Band 244 (1970).

[26] E. P. SHETTLE and J. A. WEINMAN, *Transfer of Near-Ultraviolet Irradiances Through Smog Over Los Angeles*, in: *Preprints of Papers; Conference on Air Pollution Meteorology* (American Meteorological Society, Boston, Mass. 1971), p. 121.

[27] R. J. CHARLSON, H. HARRISON and G. WITT, Science *175*, 95 (1972).

[28] S. I. RASOOL and S. H. SCHNIEDER, Science *175*, 96 (1972).

[29] E. LORENZ, *The Nature and Theory of the General Circulation of the Atmosphere* (World Meteorological Society, Geneva 1967).

[30] S. H. SCHNEIDER, *A Comparison of Some Recent Numerical Models of the Effects of Aerosols on the Global Climate* (unpublished private communication to J. Murray Mitchell, Jr., 1971).

[31] M. I. BUDYKO, *Climate and Life* (Hydrological Publishing House, Leningrad 1971); also see Ref. [19].

[32] W. D. SELLERS, J. appl. Met. *8*, 392 (1969).

[33] S. H. SCHNEIDER and T. GAL-CHEN, *Numerical Experiments in Climate Stability*,
      J. Geophys. Res. (to appear 1973).
[34] W. L. DONN and M. EWING, Met. Monogr. *8*, 1004 (1968).
[35] S. H. SCHNEIDER, J. atmos. Sci. *29*, 1413 (1972).

The Moderate- and Long-Range Transport of Air Pollutants and Their Removal from the Atmosphere

by D. M. WHELPDALE and R. W. SHAW
Atmospheric Environment Service, Toronto, Ontario, Canada

Abstract

The topics of moderate- and long-range transport of air pollutants and the removal of these pollutants from the atmosphere are discussed within the context of the cycling of contaminants through the ecosphere. Experimental results pertinent to three aspects of these problems are presented. Evidence is shown of the transport of sulphur dioxide over Lake Ontario for distances of up to 100 km; residence times of several hours for particulates over water are determined, and from measurements of vertical gradients of sulphur dioxide over water, which are often found to be ± 2 pphmv over 6 m, estimates of turbulent fluxes to and from the surface are made. A discussion follows which outlines approaches to further research in these areas.

Introduction

The proper functioning of our environment requires that cycling of natural substances through the ecosphere is not severely disrupted. A well-known example of a natural cycle is that of carbon, in which carbon and carbon compounds circulate amongst various reservoirs: rocks, plants, animals, the oceans and the atmosphere, through the processes of combustion, photosynthesis, decay, respiration and ingestion. When substances resulting from man's activities, whether carbon dioxide, pesticides or phosphates, become plentiful enough to affect the natural cycling of substances or even to establish new pathways, the functioning of our environment may be impaired, resulting in a deterioration of the quality of life and possibly endangering life itself. ERIKSSON [1] and BOLIN [2] have both stressed the importance of studying the entire cycles of such substances as sulphur and carbon dioxide (significant fractions of which may be man-made) and thus of studying their effect upon the total environment.

There is a growing body of evidence that man-made substances have atmospheric residence times that can last as long as days, months or even years and can span a large fraction of the earth. For instance, RODHE [3] has estimated that about half the sulphur deposited in southern Sweden is from sources on the European continent and in the United Kingdom. Conversely, a large fraction of the sulphur emitted by Swedish sources is quite clearly exported outside the country. Based on these studies, ROHDE and GRANDELL [4] conclude that atmospheric aerosols have a turnover time of from 2 to 12 days. ROBINSON and ROBBINS [5] found high concentrations of carbon monoxide (as much as 1.0 parts per million) in air over the Greenland ice cap whenever the air mass had previously passed over the Great Lakes region of North America and the northeastern United States, a source region for carbon monoxide produced by automobiles.

Pollutants can undergo chemical reactions in the atmosphere and can span more than one medium; they can pass from air to water to living substances and into the food chain. The Joint Group of Experts on the Scientific Aspects of Marine Pollution (GESAMP [6]) have estimated that 50% of the lead reaching the oceans does so via transport through the atmosphere and subsequent transfer from air to water, where it is potentially hazardous to marine life.

In research to date, much emphasis has been placed upon the study of the dispersion of air pollutants within distances of a few kilometers from their sources (e.g., urban studies and point-source plume dispersion studies, often without an awareness that this is only the beginning of the cycle. While recognizing the importance of such studies, we feel that there is need for continuing and expanding studies of medium-range (several tens of kilometers) and long-range (hundreds or thousands of kilometers) transport of air pollutants with concurrent studies of removal processes by which air pollutants are transferred to other media. Increasing global levels of carbon dioxide and particulates, acidification of lakes and soils, and accumulations of pesticides in the marine environment are only a few examples which emphasize the necessity of the regional and global approaches. In this paper, therefore, we hope to stimulate more interest in these topics by outlining some of their inherent problems and by briefly describing our own beginnings in studies of medium-range transport of sulphur dioxide and of removal processes affecting sulphur dioxide and particulates.

2   Medium- and Long-Range Transport of Air Pollutants

2.1   *Meteorological aspects*

A theoretical treatment of the dispersion of air pollutants may be found in PASQUILL [7]. Although there is some hope that, for inert pollutants under steady-state conditions over flat homogeneous surfaces, methods for calculating diffusion over short distances (PASQUILL [7]; GIFFORD [8]) can be extrapolated to moderate distances of perhaps 100 km, MUNN and BOLIN [9] point out that in most cases of medium- and long-range transport the problem is much more difficult for the following reasons:

a) The dispersive qualities of the atmosphere are no longer steady in time and space. For instance, day-time heating encourages vertical mixing while nocturnal radiational cooling discourages it.

b) The direction and speed of the wind varies with height and location. A further complication is that there is a diurnal variation in wind velocity in the lowest 1 or 2 km.

c) The roughness and temperature of the surface over which the pollutant travels is not uniform. Both characteristics affect the vertical dispersion of pollutants.

d) The pollutant may be removed from the air by interface processes (gravitational deposition or direct molecular impaction), by precipitation scavenging, or by chemical transformation. The first two mechanisms will be discussed in section 3.

It has been observed that a pollutant will be transported horizontally in measurable quantities over medium and long distances if its vertical mixing and dilution is limited. The smaller the decrease of air temperature with height, and the smoother and more uniform the underlying surface, then the less the vertical mixing will be. PETERSON [10] was able to follow argon-41 in a plume out over the Atlantic for a distance of 300 km because the relatively smooth surface and neutral buoyancy conditions restricted vertical mixing and dilution.

Temperature inversions (i.e. layers where air temperature increases with height) act as upper barriers or lids to vertical diffusion. Inversion layers may be adjacent to the ground or elevated above it. In the latter case, pollutants are often uniformly mixed in the vertical beneath the inversion and they may be transported over long distances. THOMPSON [11] tracked fluorescent particles for distances over 100 km beneath an inversion associated with subsiding air. CARLSON and PROSPERO [12] have detected large amounts of African dust arriving at the surface in Barbados in the Trade Winds. They have concluded that dust-laden Sahara air travels westward in a layer or 'duct' between heights of 1.5 and 4.5 km. Temperature inversions form the upper and lower boundaries of the duct and prevent vigorous vertical mixing of the dust. By the time the air mass had reached Barbados, however, measurable amounts of dust were present at ground level through the processes of gravitational settling and limited mixing through the bottom of the duct.

The above examples illustrate that the vertical temperature structure of the air is an important factor in medium- and long-range transport. Our study of medium-range transport, to be described in the following section, uses measurements of sulphur dioxide concentration at a single surface location, and relates them to regional wind direction and vertical gradient of temperature, The data, when averaged appropriately, provide evidence of transport of sulphur dioxide over distances of at least 100 km and indicate that the vertical gradient of temperature is an important factor in this transport.

2.2 Medium-range transport of sulphur dioxide over Lake Ontario

In this study, measurements of sulphur dioxide concentration from a lakeshore site in suburban Toronto (population 2.5 million) for the years 1969 and 1970 were examined for evidence of transport over distances of up to 100 km across the western end of Lake Ontario. Mean concentrations were compared for two cases: first, when the regional wind was blowing from the Lake, on the other side of which there were large industrial sources, and second, when the wind was from the land and the nearby sources in Toronto.

The sulphur dioxide sampler, operated by the Ontario government, was located on the top of bluffs about 60 m above the water and 100 m from the shoreline. There were no sources of sulphur dioxide between the sampler and the lakeshore. Mean daily concentrations of sulphur dioxide with an accuracy of ± 0.2 parts per hundred million by volume (pphmv) were used. The wind that was most indicative of regional transport was assumed to be that at the 850 millibar level (a height of about 1.5 km) reported at 1200 GMT by the nearest radiosonde station (Buffalo, New York) which was about 95 km south-southeast of the sampling site.

Other studies, such as those mentioned in the previous section, have indicated the importance of the vertical distribution of temperature for transport of pollutants over large distances. Unfortunately, detailed temperature data up to a height of several kilometers were not available at the sampling site. However, values of vertical temperature gradient $\Delta T/\Delta z$ between heights 10 and 91 m above ground were available from a meteorological tower located 25 km down the shoreline and about 2 km inland. The tower had surroundings and exposure similar to those at the sampling site. These data, while not ideal, gave some indication of the conditions of vertical mixing in the lowest layers of the air arriving at the sampling site.

Table 1
Mean sulphur dioxide concentrations at a lakeshore sampling site with respect to regional wind direction – 1969 and 1970.

Season	Regional wind	No. of cases	Mean SO_2 (pphmv)	Standard deviation (pphmv)	$\Delta T/\Delta z$ (°C/100 m)
Heating	from lake	27	1.8	0.8	−0.20
(Nov. to Apr.)	from land	208	3.0	0.8	−0.35
Non-heating	from lake	32	2.2	1.4	0.15
(May to Oct.)	from land	260	3.0	1.8	0.15

Table 1 shows the number of days, mean values and standard deviations of sulphur dioxide concentrations, and vertical gradients of temperature with respect to regional wind direction. The data were first divided into the non-heating season (May to October), when the sources are mainly scattered, strong point sources such as thermo-electric generating stations and industries, and the heating season (November to April), when the above sources are joined by domestic and commercial space-heating sources. The data were further divided according to whether the regional wind was blowing across the lake or from the land. The off-lake sector was carefully chosen to avoid including any nearby sources in Toronto. The results are as follows:

a) Heating season: The mean daily concentration of sulphur dioxide increased from 1.8 pphmv when the regional wind was from the lake to 3.0

pphmv when it was from the land. Because of the day-to-day variations in sulphur dioxide concentration, it was necessary to show by means of the 'Student's t-test' (PANOFSKY and BRIER [13]), that this increase was statistically significant. The increase is significant at the 1% level. Even when the regional wind was blowing from industrial sources across the lake, however, the mean daily concentration was at least half that occurring when it was blowing from nearby sources in Toronto, implying that regional transport was taking place. Further evidence of regional transport will be found when the non-heating season is examined.

b) Non-heating season: In this case, there was no statistically significant variation in mean sulphur dioxide concentration with regional wind direction, and the standard deviations were about twice as great as they were in the heating season. Of most interest to us is the finding that winds blowing from sources across the lake result, on the average, in concentrations of sulphur dioxide approximately as great as those with winds blowing from nearby sources on land. Our interpretation of the high standard deviation with the wind blowing across the lake is that on certain days, high concentrations result from regional transport. Enough of these days occur so that when the data are averaged over all cases, the importance of regional transport from sources across the lake is about as great as short-range transport from the land.

The values of temperature gradient $\Delta T/\Delta z$ (right-hand column in Table 1) are algebraically greater in the non-heating season. In particular, the positive values in that season indicate inversion conditions which are associated with limited vertical mixing and dilution of pollutants and thus favourable conditions for long-range transport. Although temperature soundings and sulphur dioxide measurements up to a height of two or three kilometers over the lake are needed for further study of this problem, it appears that the smaller number of local sources and the relatively limited vertical mixing are two important reasons for the medium-range transport being as effective as short-range transport in the non-heating season. Further study needs to be made of the role of the local wind which, due to the presence of the lake and the city, may on occasion be from a direction different than that of the regional wind.

3   The Removal of Pollutants from the Atmosphere

3.1 *Introductory remarks*

The topic of pollutant removal from the atmosphere is one deserving of substantially more investigation because it is in this area that the harmful effects of pollutants on their receptors becomes evident. Before looking in detail at some of our current research projects in this area, a few examples of the effects of air pollutants on their sinks will be outlined. A recent report, SWE-DEN [14], prepared for the United Nations Conference on the Human Environment, shows that the pH-value, which is a measure of acidity, of many of Sweden's lakes and rivers has been falling because of the deposition of sulphuric acid aerosol from anthropogenic sources of sulphur dioxide throughout Europe. If present development continues, the report goes on to say, with concurrent increases in sulphur emissions, about 50% of the lakes and rivers in the areas concerned may have, in less than 50 years, a pH-value of 5.5 or 5.0; this will be critical for the surviral of most fish. A similar assessment of these effects on some forested areas of Scandinavia indicates that an annual rate of reduction of the growth rate of about 0.3% is occurring.

A second example of the impact of atmospheric pollutants on sinks, also in the form of a contribution of air pollution to water pollution, is that of the highly toxic, heavy metal, lead. The annual world production of lead is approximately 3.5×10^6 tons; more than 10% of this is used in motor fuels and subsequently released into the atmosphere from exhausts in aerosol form. It is estimated (GESAMP [6]) that as much as 2×10^5 tons of lead is introduced into the sea by precipitation scavenging of this aerosol. As a result, lead concentrations in surface waters of oceans in the northern hemisphere have increased from about 0.01 to 0.02 µg/kg to about 0.7 µg/kg, despite the rather short residence time for lead of about 2 years.

There are three main sinks for atmospheric pollutants: precipitation scavenging, chemical reactions in the atmosphere and depletion at the atmosphere-earth interface. The removal of both gaseous and particulate pollutants by precipitation is the result of in-cloud scavenging called rainout and snowout and of below-cloud scavenging called washout. Although the numerous microphysical processes which comprise scavenging are becoming better understood, we are still lacking a quantitative picture of precipitation scavenging on the regional and global scales.

Many pollutants participate in chemical reactions in the atmosphere; for example, sulphur dioxide is converted to sulphuric acid in the presence of fog. Although chemical reactions can thus be thought of as sinks in the cycling of some substances, one must be aware that they create new substances and thus act as sources for other pollutants. The field of atmospheric chemistry is an immense one in which there is much activity today.

The third sink for pollutants is the one of most interest to us – that of depletion at the atmosphere-earth interface. Gases and submicron particulates are brought to the interface by turbulent transfer and larger particulates by gravitational settling. The efficiency of these processes is one of the factors controlling the exchange. The other factor is the effectiveness of the surface, whether vegetation, soil or water, in retaining the pollutants which arrive there.

We feel that one of the most pressing requirements in the area of pollutant sinks is for parameterization of pollutant removal over moderate to large areas. It is necessary to know at what rate lakes, oceans, forests and soils are being polluted from air-borne sources. Specifically, the two areas which we shall examine are, first, the residence time of particulates over inland lakes on a regional scale and an estimate of quantities deposited, and secondly, an examination of sulphur dioxide gradients over water, and of their turbulent fluxes, so that estimates of the exchanges rates may be made.

3.2 *Particulate residence times*

The study of particulate residence times over water employed data collected during an international program of air pollution investigation in an area of the Great Lakes Region of North America. The program was established to investigate complaints of transboundary flows of pollutants in two areas, separated by about 85 km, Detroit, Michigan – Windsor, Ontario and Port Huron, Michigan – Sarnia, Ontario. A detailed description of the geography of the area, measuring program and source emission inventory can be found in the report of the International Joint Commission (IJC [15]). Among the data available were wind speed and direction measured at approximately 20 stations and 24-hr suspended particulate concentrations measured at approximately 50 stations spread throughout the region, an aera of more than 9,000 km². Two large lakes, Huron and Erie are partially in this area and Lake St. Clair is completely within it. The following criteria were used to select cases (24-hr periods) for the study of particulate residence times:

1. Sufficient data were available to establish a regional background value of particulate concentration;

2. a lakeshore station was available such that winds were from off the water at the station of the 24-hr period of high volume samples monitoring.

Table 2
Summary of data for determination of particulate residence times.

Date	Lake	Suspended particulate concentration (μg m^{-3}) regional background	lakeshore station	Mean wind speed (m s^{-1})	Distance over water (km)	Precipitation during 24-hr period
11 May 1968	St. Clair	54	43	4.5	28	yes
24 May 1968	St. Clair	75	54	5.3	32	no
26 May 1968	St. Clair	38	28	5.3	30	yes
16 June 1968	St. Clair	34	28	3.6	28	yes
10 July 1968	Huron	24	12	5.3	200	no
25 July 1968	Huron	26	10	3.6	200	no
20 Sept. 1968	Erie	52	39	4.0	50	yes

During the 14-month period of the study, there were 7 days which fulfilled these two criteria. The cases are summarized in Table 2. Although four of these seven days had a measurable amount of precipitation, these cases are not noticeably different. Concentrations of suspended particulate are found to be substantially lower than regional background values whenever the flow is over a moderate stretch of open water. Assuming an exponential sink term over water, of the form

$$c(t) = c_0 e^{-t/\tau}$$

where $c(t)$ is the concentration of suspended particulate after a travel time t over water from an initial location where the concentration is c_0, the residence time, τ, for the suspended particulate may be determined. From the seven cases in Table 2, a mean value of $= 10.4$ hours was found, with a standard error of ± 4.2 hours. That is, for every 10 hours of travel over water, the concentration of suspended particulate is deplected by a factor of e^{-1}. This appears to be a reasonable value for particulate residence time in such a situation since MUNN [16] found values of the order of 2 hours for dispersion from an urban source where particles would be somewhat larger and thus would fall out more quickly. Tropospheric aerosols which would generally consist of smaller particles than those in our study (where there are industrial sources in the region) typically have residence times of the order of a few days to a few weeks JUNGE [17].

To estimate the average depletion rate of particles over a 24-hr period, consider case 2 in Table 2, i.2., 24 May 68. We shall assume that the lake has a characteristic dimension $L = 40$ km, that there is a constant uniform wind $\bar{u} = 5.3$ m s^{-1}, and that the vertical distribution of suspended particulate is uniform through a mixed surface layer of height $H = 500$ m (MUNN [16]). The mass of material depleted by the lake is then

$$HL\bar{u}c_0 \,(1-e^{-L/\bar{u}\tau})\,(24)\,(3{,}600) \ \mu g$$

or approximately 200 tons. The average flux into the lake is therefore

$$\frac{H\bar{u}c_0\,(1-e^{-L/\bar{u}\tau})}{L} \ \mu g \ m^{-2} \ s^{-1}$$

or about 1 µg m^{-2} s^{-1} during such a period.

The value of such work is that estimates can be made of residence times, and thus of the efficiency of sink processes. This will contribute to a better understanding of the efficiency of the cleansing processes in the atmosphere and the strength of pollutant sinks.

3.3 *Vertical gradients and turbulent fluxes of sulphur dioxide*

The second project, dealing with the turbulent transfer of pollutants, in this instance sulphur dioxide, is a program within the International Field Year on the Great Lakes (IFYGL). This is a binational (Canada, United States), multidisciplinary, comprehensive study of Lake Ontario and its watershed. The concentration of sulphur dioxide is being continuously monitored during intensive periods at two levels, nominally 2 and 8 m, over the lake from an anchored platform approximately 2 km off shore in the western end of Lake Ontario. The sulphur dioxide measurements are being made concurrently with those of turbulent fluxes of momentum, heat and moisture so that when all data are available from the Field Year, fluxes of sulphur dioxide can be estimated assuming equivalence of turbulent diffucivities. At present we shall restrict ourselves to describing examples of gradients measured. Sulphur dioxide can be detected when the wind is blowing from the direction of any of three large industrial centres, Toronto, Hamilton or Buffalo, all 50–60 km distant. The lower limit of detection is 0.1 pphmv.

When measurable concentrations exist, there is often a gradient between the two measuring levels. The figure shows records of two days when gradients existed. Positive values in the figure indicate higher concentrations at the 8-m level, i.e., a flux of sulphur dioxide toward the water. Gradients as small as 0.2 pphmv per 6 m were measurable, as indicated by the dashed lines in the figure. These two sample records of sulphur dioxide gradients demonstrate the following: first, gradients do exists over a water surface; second, this gradient can change sign over periods of less than an hour; and third, values of the gradient of up to 2 pphmv per 6 m are not unusual. Meteorological variables such as wind speed and direction, temperature gradients, as well as surface water conditions such as temperature, pH and chemical composition undoubtedly control the magnitude and direction of these sulphur dioxide gradients. In the future, such intensive supporting observations will be made.

Our measurements of sulphur dioxide gradients enable us to estimate turbulent fluxes and thus pollutant exchanges at the inteface. For the examples shown, a turbulent diffusivity, K, for sulphur dioxide can be assumed to be

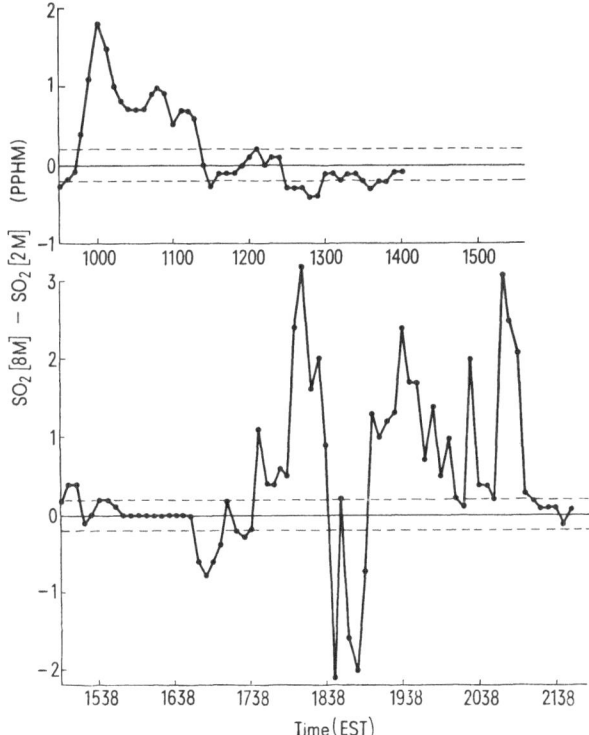

Two sample records of sulphur dioxide gradients between the 8- and 2-m level

equal to that for water vapour, a representative value of which in this situation is 10 m^2 s^{-1}. Using the relation for the flux, F, of sulphur dioxide,

$$F = K \, \Delta[SO_2]/\Delta z,$$

where $\Delta[SO_2]/\Delta z$ is the concentration gradient of sulphur dioxide, fluxes of the order of ± 10 to ± 100 μg m^{-2} s^{-1} are found to be occurring.

Once such fluxes have been measured under a variety of atmospheric and surface conditions, then the exchanges of pollutants at the interface can be parameterized on a regional basis. Then it will be possible to reliably estimate pollutant sink strengths.

4   Conclusions

The three studies described above have provided us with a better insight into some areas of transport and removal of pollutants. We have been able to find evidence of medium-range transport and determine that a governing factor is the vertical temperature gradient. It has also been possible to determine particulate residence times and estimate sink strengths on a regional basis. Both these projects used data gathered primarily for the purposes of air quality monitoring. In the case of sulphur dioxide measurements, we determined that gradients of this pollutant can be measured over water. Turbulent fluxes can, therefore, be estimated.

The consequences of this work are twofold: first, in the case of transport; since moderate-range transport can be detected, it should be possible to make quantitative estimates of amounts involved, and also to identify distant sources. Secondly, for removal studies, our results lead us to believe that it is possible to estimate fluxes and thus source strengths on a regional basis. In both cases we have learned more about the cycle of the substance involved.

In future research, some of the limitations of our work must be removed. For example, it will be necessary to obtain better spatial and temporal resolution. For instance, instead of daily averages at a single point, measurements (perhaps by aircraft) of pollutants in two or three dimensions over shorter periods of time should be made. Future experiments should be done in such a way that quantitative estimates of vertical and horizontal fluxes can be confidently made. This will require supplementary information about meteorological conditions and surface characteristics. And finally, these experiments must be done over a variety of surfaces and under a variety of atmospheric conditions. Only in this way will we be able to parameterize such processes over moderate and large scales. Because of the wide variety of processes involved, and the large space and time scales, such work will require the international cooperation of scientists of many disciplines.

References

[1] E. ERIKSSON, J. geophys. Res. *68*, 4001 (1963).

[2] B. BOLIN, *Meteorological Challenges: A History* (Information Canada, Ottawa 1972).

[3] H. RODHE, Tellus *24*, 128 (1972).

[4] H. RODHE and J. GRANDELL, Rep. AC-19, Int. Inst. Meteorol., University of Stockholm, 33 p. (1972).

[5] E. ROBINSON and R. C. ROBBINS, J. geophys. Res. *74*, 1968 (1969).

[6] GESAMP, IMCO/FAO/UNESCO/WMO/IAEA/UN, Joint group of experts on the scientific aspects of marine pollution. Report of the third session (FAO Headquarters, Rome 1971), 9p. + 9 annexes.

[7] F. PASQUILL, *Atmospheric Diffusion* (Van Nostrand and Co., London 1962, ix + 297p.)

[8] F. A. GIFFORD, Nucl. Saf. *4*, 47 (1961).

[9] R. E. MUNN and B. BOLIN, Atmos. Env. *5*, 363 (1971).

[10] K. R. PETERSON, J. appl. Met. *7*, 217 (1968).

[11] N. THOMPSON, Met. Mag. *93*, 193 (1964).

[12] T. N. CARLSON and J. M. PROSPERO, J. appl. Met. *11*, 283 (1972).

[13] H. A. PANOFSKY and G. W. BRIER, *Some Applications of Statistics to Meteorology.* (Pennsylvania State U., University Park, Pennsylvania 1963), ix + 224 p.

[14] SWEDEN, Air Pollution across national boundaries; the impact on the environment of sulphur in air and precipitation. Sewden's case study for the United Nations Conference on the Human Environment (Royal Ministry for Foreign Affairs, Royal Ministry of Agriculture, Stockholm 1971), 130 p.

[15] IJC, Joint air pollution study of St. Clair – Detroit River areas (International Joint Commission, Ottawa and Washington 1971), 241 p.

[16] R. E. MUNN, Paper presented to International Geographical Congress, Montreal, Canada (1972).

[17] C. E. JUNGE, *Air Chemistry and Radioactivity* (Academic Press, New York and London 1963), 382 p.

Effects of Air Pollution on Forest Vegetation under Northern Conditions

by Satu Huttunen
Department of Botany, University of Oulu, Oulu, Finland

Abstract

This paper describes briefly some methodological investigations of tree damage caused by air pollution in northern Finland. The methods presented are dendrochronology, applied to the trees of the damaged area, and microanalysis, applied to demonstrating the presence of sulphur and other pollutants in a biological material. These biological parameters are important in investigating the pollution of air in northern regions, where the cold climate makes nature less resistant than in areas with more favourable conditions. The article also contains some references to the present stage and the future prospects of plant-ecological research.

1 Introduction

Pollution has brought numerous large-scale global problems into the area of biological research. The cold northern climate, which in itself constitutes a restriction on various forms of life, makes the northern nature, which is generally considered so rugged and hardy, more sensitive to changes and more vulnerable in many respects than the southern formations with more favourable conditions. The northern coniferous forest zone, upon which the economies of several countries largely depend, is the element most severely affected by air pollution.

Research into air pollution was slow to begin, particularly in northern Europe. It was tenaciously maintained that the large intact areas guaranteed the purity and adequacy of nature, and the alarming messages from other parts of the world were unable to shake this notion. Several local catastrophes of noticeable proportions were required before the attention of the general public and research workers was drawn to this point.

The productive capacity of northern vegetation is greatly limited, and any small adverse change may crucially disturb the functioning of ecosystems and thereby upset the equilibrium. The quantity of primary production under northern conditions has been under investigation for several years (e.g., in the IBP project). But the changes which may already have taken place have not been studied on a larger scale. The new Man and Biosphere project will continue productivity studies with special regard to the possible changes and the development of different biological monitoring techniques. One of the urgent problems at present is to find biological parameters for identifying the appearance of global and local effects of air pollution. Investigations to this end have already been started, and various suitable parameters have been found.

The importance of biological parameters as a supplementary technique to the mechanical and physical methods of measurements lies in the very fact that many biological organisms react to environmental changes before any measurable signs appear, and changes we are today capable to measure are often so serious that damage to nature has already occurred.

The monitoring and investigation of coniferous forests, in particular, is one of the most important aspects of the northern studies. The IBP program in Finland has included investigations of *Hylocomium-Myrtillus*-type forests with the spruce (*Picea abies* L.) as the dominant tree, but the scope of the studies will be extended to include dry heath type forests with the pine (*Pinus sylvestris* L.) as the dominant tree. The economic significance of dry heath type forests is also great, which is why it is important to investigate them.

As much as 400,000 hectares of forest has already been destroyed in various European catastrophes. A large proportion of this area is in Scandinavia. Most of the damage has been due to high sulphur and fluorine content in the air. The effects of sulphur on nature are both primary, i.e. toxic and secondary (sulphur reduces the pH of rain water and consequently of the soil). Rain water and winds may carry air-borne toxins over great distances. Fall-out of sulphur from the industrial areas of Central Europe has been noted in southern Sweden and Finland.

The sulphur problem is the most serious as about 50% of atmospheric sulphur is a product of human action. The proportion of sulphur produced by man continues to increase, which means that the total sulphur production also increases year by year. The situation therefore becomes progressively worse. Furthermore numerous other substances are discharged into air in great quantities, including nitrogen oxides, fluorine, chlorine, ozone, aerosols and other chemical compounds, as well as soot and various dusts, all of these affecting nature, both individually and synergistically.

The photochemical reaction of hydrocarbons and nitrogen oxides produces ozone and PAN. Internal combustion engines and industrial processes are among the major sources of hydrocarbons and nitrogen oxides. Ozone has a detrimental effect on many forms of plant life. Peroxyacetyl nitrate, commonly called PAN, is highly toxic to many plant species, especially small plants and young leaves. Nitrogen dioxide causes direct vegetational damage, and sulphur dioxide brings about acute and chronic injury. The synergistic action of ozone and sulphur dioxide may cause more severe damage than either of the pollutants does alone. Even low concentrations of fluorine damage plants. Chlorine damage occurs close to the source of pollution. Ethylene injures vegetation in urban areas. The above list of main pollutants could still be continued by adding several others, such as lead and heavy metals, which have been found to accumulate in plants. Lead does not bring about direct vegetational damage, but it accumulates in plants, thereby entering the biological nutrient cycle.

In addition to visible vegetational damage, air pollution may also bring about a decline of growth and a decrease of the total productivity of nature. The vegetational damage effected by air pollution is not only important as an economic loss to agriculture and forestry and an annihilation of natural values; it is also an important warning of the threatening problems of air pollution which may affect man and his well-being. Plants are regarded as sensitive indicators which reveal even small contents of toxic agents in the air. Some species of plants are more sensitive than others, and it is therefore exceedingly important to try and find the different indicator species. Lichens make up one sensitive indicator group. Investigations of lichens have been fairly numerous, particularly in northern conditions, where lichens are also of economic significance, and these investigations have shown the value and practicability of lichens for monitoring purposes.

It has already been pointed out that coniferous trees are sensitive to most air-borne pollutants – coniferous forests have been damaged by air pollution in different parts of the world. Thus coniferous trees are also sensitive indicators, and they have been investigated for some time with fairly good results. The following is an account some studies carried out on *Pinus sylvestris*.

2   Investigations of Tree Damage Caused by Air Pollution

2.1 *Preliminary investigations*

Under northern climatic conditions, the most severe damage is inflicted in the winter, particularly in coniferous forests. The spring of 1966 was the first time that large-scale damage to trees, particularly pines and birches, was noted in the vicinity of the chemical processing plant located at Oulu (65° N) (mixture-type fertilizers, etc.; the most northern in the world). From that year onward the damage has increased; new injuries were most numerous in the spring of 1969 and 1970. The present damaged area comprises 250 hectares, part of which has been completely destroyed. Following the spring of 1969 extensive investigations were launched in order to determine the nature of the damage.

Indicator species are relatively few in the vicinity of this industrial site, but the symptoms of the pines and birches showed that the damage had began under winter conditions. Since the gaseous exchange in trees in winter time is very slight under the local conditions, the possibility of damage due to factors other than gases suggested itself. In the early stages of research the attention of the workers was already drawn to the fall-out of fertilizer dust on the surroundings of the industrial site. Analyses revealed the composition of this dust to be the same as that of the fertilizers produced by the industrial process. It was further noted that various toxins, such as fluorine, sulphur and chlorine, were deposited on the surface of the plants along with the fertilizer dust.

The investigations showed the toxic effect to be the crucial factor. Water deficiency and the generally hard conditions prevailing in the winter enhance the sensitivity of trees to toxins. The presence of fertilizers may have diverse effects on the tolerance against toxins. Fertilizers reduce the xeromorphism of the needles, thereby making them more vulnerable to deterioration through water deficiency. The preliminary studies established fluorine as the main cause of the tree damage. The particular sensitivity of coniferous trees to fluorine has already been known for a long time. The preliminary research showed further that sulphur must be an important contributing factor in effecting the damage. The role of chlorine was not noticed until later.

The preliminary research, which was specially designed to ascertain the causes of the catastrophe, consisted mainly of different chemical analyses on pine needles and observations on the winter time water economy of the needles both in trees growing in natural conditions and in test trees brought near the industrial site. Further measurements were made on the photosynthesis and respiration of the pine, and the chlorophyll content was also observed.

The chemical analyses clearly showed that fluorine accumulated in the test trees within a very short period of time. The same phenomenon has subsequently been noted in deciduous trees on the test field near the industrial site. Observation of the water content showed that the critical point was reached long before the beginning of spring. Even unpolluted trees frequently approach the critical level at that time of the year. The photosynthesis and respiration of coniferous trees showed features suggestive of severe disorders in the physiological equilibrium of the trees [1].

The investigation of photosynthesis seemed to suggest that changes take place in the physiology of trees long before actual visible damage appears, and that these changes effect growth.

2.2 Dendrochronological investigations

The radial growth of trees during the last twenty years was investigated in the damaged area, and the values obtained were compared with the corresponding values of trees growing in unpolluted conditions.

The work was primarily a methodological experiment, applying dendrochronology to the study of the effects of air pollution on the radial growth of pines in the surroundings of the said industrial establishment. The production of fertilizers has been going on for about 20 years, and fertilizer dust and toxins, particularly fluorine compounds, have been emitted in the process. The radial growth of pines clearly displayed variations which correlated with the industrial activity. The fairly young trees (under 50 years old) were especially sensitive in their reactions to the fertilizing action. The detrimental effects do not appear until later; they are also visible in older trees (under 100 years old). The variation of radial growth can be regarded as a biological parameter which should be taken into account when investigating the effects of pollution on coniferous forests and forecasting the future development.

Young trees showed the effect of fertilizers fairly soon after the initiation of industrial production. An increase of radial growth first appeared after the year 1952, when the production of saltpeter was started. The production of mixture-type fertilizers began in 1957, and apatite containing fluorine was then

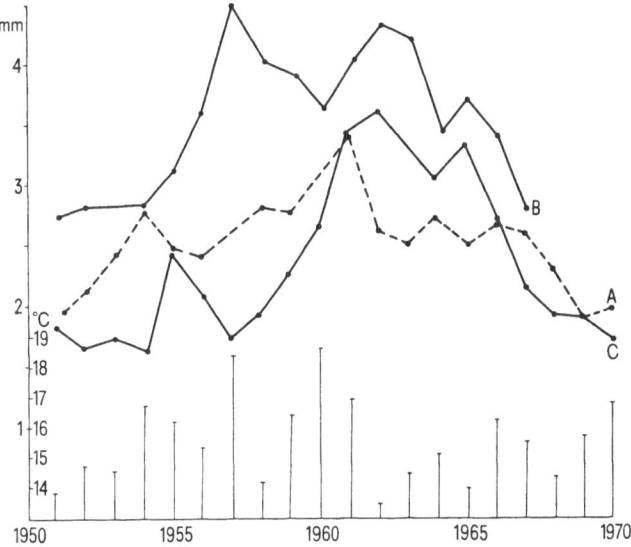

Figure 1

Width of growth-rings in trees under 30 years old and mean July temperatures
(vertical lines) in Oulu during the last 20 years. *A* trees (No. 4) growing in the unpolluted
area; *B* trees (No. 5) growing on the industrial site (Typpi Oy); *C* trees (No. 3) growing
at some distance (0.5–1.5 km) from the industrial plant.

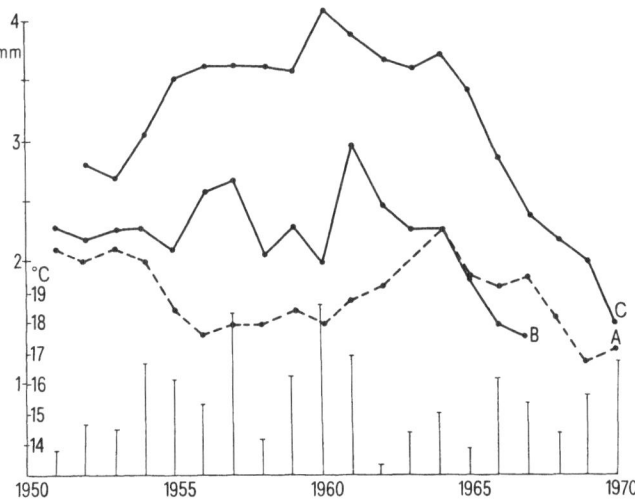

Figure 2

Width of growth-rings in trees 31–55 years old and mean July temperatures
(vertical lines) in Oulu during the last 20 years. *A* trees (No. 3) growing in the unpolluted
area; *B* trees (No. 4) growing on the industrial site (Typpi Oy); *C* trees (No. 7)
growing at some distance (0.5–1.5 km) from the industrial plant.

introduced as a raw material. A remarkable decrease of radial growth became apparent from the early 1960's onwards. The decrease was characteristic of trees under 55 years old. Trees older than 55 years, on the other hand, did not react to the fertilizing effect in any way; their radial growth began to decrease steadily from the early 1960's onwards, and continued so rapidly that the final decrease of growth was a perfectly clear phenomenon. Trees over 100 years old did not react at all, i.e. they showed no changes of radial growth prior to their death. Even the normal annual growth of trees this age is very slight in the local climate conditions.

This one study seems to establish the dendrochronological method as applicable to investigations of air pollution, at least when the effects of pollutants are as strong as they were in the present case [2].

2.3 *The toxins contained in pine needles as assayed by a microanalyzer*

(Electron probe X-ray microanalyzer JXA-3 SM JEOL Japan)

The preliminary investigations carried out in the winter 1969–70 included various chemical analyses on the needles of pines growing in the damaged area. In addition to the quantities of toxins present in the needles, rates of accumulation were also determined. The chemical analyses were mainly focussed on fluorine, sulphur, and nitrogen compounds. The accumulation of fluorine into needles and the large quantities of it present in them were clearly demonstrable. The accumulation of sulphur was not quite so obvious. The preliminary investigations thus provided no clear evidence of the role of sulphur. It appeared probable however, that the role of sulphur in inflicting the injury was less important than that of fluorine, but that sulphur did act as a contributing factor [1].

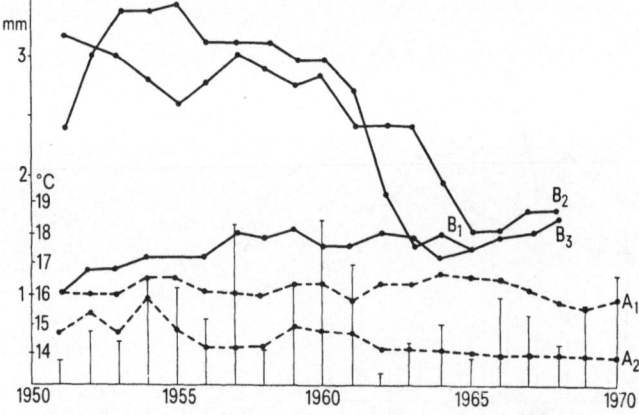

Figure 3

Width of growth-rings in trees over 55 years old and mean July temperatures (vertical lines) in Oulu during the last 20 years. A_1, A_2 trees growing in the unpolluted area: A_1 71–100 years old (No. 2); A_2 over 100 years old (No. 2). B_1, B_2 trees growing on the industrial site (Typpi Oy): B_1 56–70 years old (No. 2); B_2 71–100 years old (No. 8); B_3 over 100 years old (No. 4).

As the work was carried further, the effects of sulphur were elucidated by the method of microanalysis. The application of this method to biological materials, particularly plants, has hardly ever been tried before, and most of the present work therefore consisted of the development of techniques. This method has certain great advantages:

Very little material is needed for the assays: just a few needles from the trees under study suffice for the analysis. Moreover, the material can be prepared easily by a cryostat or manually.

The needles were cut into cross section preparations, which were analyzed for sulphur and chlorine by means of the line analysis. The preparations were obtained from the middle of the needle, and the run was performed in such a way that the central cylinder was traversed. The samples of needles collected from the damaged area were compared with samples obtained from unpolluted trees. The microanalyses revealed a fairly obvious difference between the polluted and the unpolluted samples. The analyses of unpolluted needles showed a highly uniform intensity in all tissues. Sulphur contents of unpolluted trees were very low or non-existent. The normal sulphur content of pine needles according to chemical assays is 0.1–0.2% of dry weight [3, 4]. However, the intensity of microanalysis is probably not in itself sufficient for estimating the sulphur content, since the device cannot be calibrated for any known sulphur content in a similar material when tissue materials are used.

Numerous comparative analyses of samples obtained from more and less polluted nature will probably reveal a general trend which will provide fairly ample information on the toxins contained in the tissues of the needles. The chlorine analyses of unpolluted needles were consistently at the O-level, but the samples from the polluted area clearly appeared to contain chlorine compounds.

The analyses were compared by marking the intensity of the analyses of unpolluted samples as 1, and presenting the intensities of the polluted samples as multiples of this. The analyses of damaged needles showed intensities even tenfold in comparison with the standard. Photographs taken of the samples made it possible to locate the peaks representing elevated intensity in the different types of tissues. – The first highly interesting observation was made at this stage: in nearly all of the damaged samples the peaks were located in the transfusion tissue region at both the upper and the lower side of the needle. Occasional peaks were also noted in the assimilation parenchyma, i.e. in photosynthesizing tissue. The height and the number of the peaks were nearly invariably proportional to the condition and the degree of injury of the needle. The more severely damaged the needle was, the easier it was to indicate its sulphur and chlorine content by means of microanalysis.

When more analyses were made, it was possible to indicate undisputable sulphur and chlorine contents even in samples obtained from the margins of damaged area, where the needles did not yet show any visible symptoms.

The method of microanalysis has, however, not yet undergone sufficient testing, and it is therefore too early to say whether this method can be used to determine clearly the entrance of toxins into the needles before any visual damage or chemically measurable changes appear. But it can be said with certainty that the changes which are detectable visually can also be shown by microanalysis, and the translocation of toxins within the needles can be indicated at least in the different types of tissue, occasionally even in individual cells.

In interpreting the analyses, attemps were made to eliminate all the possible sources of error, such as differences in the O-level on the different occasions, differences due to the unevenness of the sample, etc. The analyses were frequently done with bilateral electron beams, i.e. the run was carried out on both sides, in order to eliminate the changes of intensity due to the unevenness of the sample.

The conclusions presented above are based on more than fifty individual assays, each of which was interpreted separately, independent of the others, and the results of which generally remained similar troughout the experiment. It is natural, of course, that there is great dispersion in the results of the different samples due to various factors, but the results can probably be regarded as a rough indication of the general trend.

In my personal opinion the best point of microanalysis lies in the fact that even a brief visual survey provides information on the translocation of different substances. The variation of intensity can in itself probably be used as a quantitative standard of some kind, if one bears in mind that a sevenfold intensity

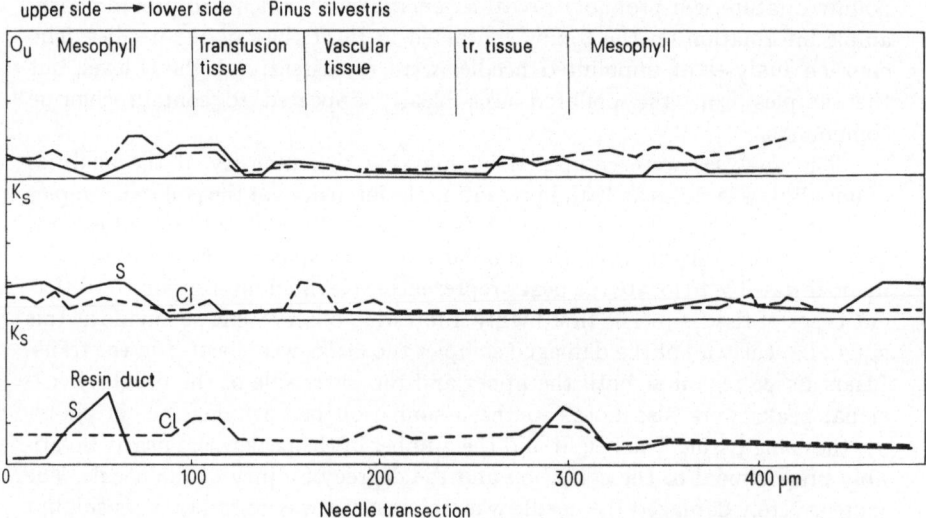

Figure 4

Cross section of a needle of *Pinus sylvestris*. Standard samples obtained from the unpolluted area. The intensity in these samples is low and relatively even. The samples from the polluted area were compared with these samples separately each time.
Dash line: chlorine intensity; continuous line: sulphur intensity. The samples were obtained from two-year-old needles. (SK α mica 100 cps, Cl stearate 100 cps, specimen 10 μm/min, 15 kV; University of Oulu, Institute of Electron Optics.)

cannot be taken to represent a sevenfold quantity of the substance in comparison with the unpolluted samples, though it can certainly be regarded as a clearly elevated content of the substance in question.

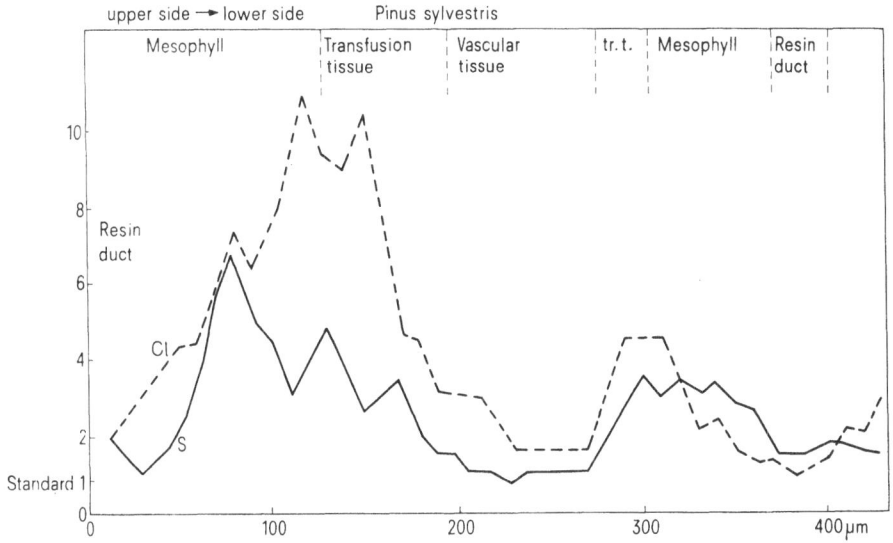

Figure 5

Exceedingly high chlorine intensity in the sample, particularly in the tissues of the upper side. Sulphur intensity is also remarkably high. The sample was obtained from the most severely damaged area in May 1971. The needle under analysis shows symptoms typical of fluorine damage. The analysis was made on the green parts (approximately at the middle) of two-year-old needles.

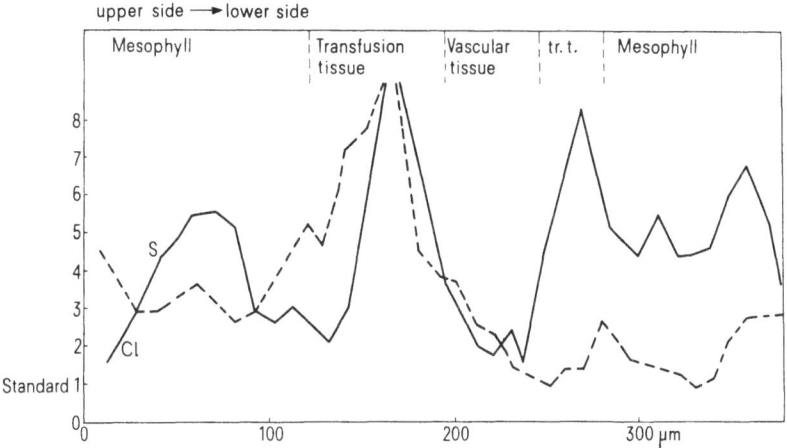

Figure 6

High sulphur intensity at both the upper and the lower side. The sample was obtained from an open area in the middle of the most severely damaged area.

The final location of the different toxins in the leaves of plants and the needles of coniferous trees has a great influence on the appearance and the degree of severity of the injury. Microanalysis is doubtless a suitable method for analyzing sulphur, chlorine and heavy metals. So far we have not been able to try microanalysis on fluorine compounds, because we have not had the additional device of the microanalyzer for reaching the lighter elements at our disposal. We shall probably be able to try fluorine assays in the near future.

2.4 *Future prospects of research*

We do not yet know enough about the effects of air-borne pollutants on the productive capacity of terrestrial ecosystems and their different sub-systems. The purpose of the research is to continue elucidation of energy and nutrition currents, in order to make the functional models of the ecosystems, on which work has already been started, sufficiently reliable, and to enable the initiation of a survey of changes (e.g., the HMT ecosystem in the forest zone). The work on ecosystems will now be expanded to include the dry pine-dominated ecosystem at the individual, population and ecosystem levels. At the individual and population levels, research will be made to cover an ever increasing number of species: lichens, trees, and dwarf shrubs, which can be used as means of widening the scope of research to include arctic regions. Continuous, effective and versatile observation of the pollution situation will be necessary in the future.

The future tasks of research also include that of finding the species of trees which are resistant against pollutants, even under strenuous climatic conditions. Further, it is important to investigate the interindividual differences observable within the species. The elucidation of the provinience problems of main species of trees constitutes another major object of study. This concerns especially *Pinus sylvestris*, *Picea abies* as well as *Betula verrucosa* and *B. pubescens*. These species make up most of the forests in Scandinavia, and investigation of the different proviniences is therefore also economically important.

If air pollution is regarded from the point of view of single pollutants and the information available on the different polluting agents, it must be concluded that sulphur continues to be the most important object of study, despite the fact that a fair amount of information concerning sulphur and its effects has already been compiled. The various substances present in exhaust gases, e.g. lead, will play an increasing role in nature, and it is therefore important to investigate them. A number of different substances were enumerated above which are generally known to have detrimental effects, but not much is known about their effects in the northern climate, and the subject thus calls for further study. In addition to all this, there are several special problems, occasionally of great local significance, which have not been studied beyond the very elementary stages. There are also some relatively new chemical compounds which are being used in increasing quantities. It is important to study them all together and separately, and to study them in those particular areas where the different forms of life are most restricted and the climate creates an easily unbalanced ecosystem. Personally, I regard the observation of arctic regions as important, since the effects of pollution may be discernible even more easily there than in forest-grown areas.

The search for new indicators must not surpass in importance the search for species which would be suitable for repairing the damage already inflicted and could be used as protective plantations in industrialized and greatly urbanized areas. A pleasant cultural environment which is well attended will save nature from excessive and inconsiderate utilization.

3   Conclusion

It is my personal conviction that more and more biological monitoring and measuring methods, i.e. parameters which can be used to detect the possible changes earlier, will be needed in the future. Some factors which are now considered quite harmless to nature may turn out to be detrimental at some later stage. The tolerance and adaptability of nature in regard to this one factor may be changed by other factors in the course of time and in such a case monitoring is essential. It will also be useful to have at hand research methods for testing substances which will possibly be invented and introduced into use.

When a biologist discusses air pollution and its effects on nature and its ecosystems, he always adheres to the basic principle that the purpose of research and knowledge is to secure the opportunities of man to enjoy an unpolluted environment and to live on the earth. It seems that the task of biological research on air pollution under northern conditions is to concentrate on the special characteristics and ecosystems of northern climate. It is also important that the work be carried out here and now; tomorrow the changes may already be so far advanced that the original or even near-original nature of the north cannot be investigated, let alone saved.

References

[1]   P. J. Havas, Acta For. Fenn. *121* (1971).

[2]   P. J. Havas and S. Huttunen, Biol. Conserv., vol. 4, No. 5. (1972).

[3]   M. Katz and A. W. Mc Callum, in: Mc Cabe, *Air Pollution* (1952).

[4]   E. Björkman, Stud. For. Suec. *78* (1970).

Air Pollution and Traffic Safety

by PAVLE TODOROVIĆ
Institute of Public Health, Belgrade, Yugoslavia

Abstract

Today's traffic requires from the driver an increased psychophysical capability. It is therefore most important to ensure optimal observation, failure-free transmission of impulses to the center of the cerebral cortex and subcortical ganglia formation of an adequate response and perfect transmission of the nerve impulse to the periphery, i.e. to stimulate adequate reactions by means of nervomuscular regulation.

Carbon monoxide and nitrogen oxides from exhaust gases are most prominent for their unfavorable, acute affects on this psychophysical system and have thus a direct influence on traffic safety.

The influence on driving capabilities of CO and NO_x concentrations measured at some crossroads in Belgrade and inside cars are studied in this report.

1 Introduction

The air pollutants originating from motor vehicles are of special importance, as they have, besides their longlasting effects (lead, carbohydrates) very distinct *acute* effects on the exposed population.

Persons professionally exposed to traffic effects, such as taxi and truck drivers, traffic policemen, newspaper sellers, are especially endangered by harmful traffic effects, as are people living near busy roads.

It is of absolute necessity that drivers have an adequate psychophysical capability for driving in order to ensure safe driving in modern traffic. It is, therefore, most important to assure optimal observation, failure-free transmission of impulses to the center in the cerebral cortex and subcortical ganglia, formation of an adequate response and perfect transmission of the nerve impulse towards periphery, i.e. to stimulate adequate reactions by means of the nervomuscular regulation. This complex of actions is regulated by nerval, hormonal and humoral mechanisms, where the condition of the nervous system has major influence.

It is therefore important for safe driving to take care of the state of the nervous system – especially the central nervous system – to secure optimal functioning of the nerve cell.

Carbon monoxide and nitrogen oxides are most prominent for their unfavorable, acute effect and exhaust gases produced by motor vehicles have thus a direct influence on traffic safety.

Carbon monoxide is a well-known blood poison which has an exceptionally great affinity, about 300 times greater than oxygen, for combining with the blood hemoglobin.

The quantity of oxyhemoglobin is reduced by combining with hemoglobin and by forming carboxyhemoglobin. At the same time the quantity of oxygen, transported by means of hemoglobin to the cells, especially nerve cells, is reduced. Similar effects are found in the group of nitrogen oxides which achieve their effect by combining themselves with hemoglobin, transforming it into methemoglobin, and by so doing lessening the possibility of transporting appropriate oxygen quantities to the cells.

From the aspect of acute affects, carbon monoxide is far more important, for it shows, besides outstanding affinity to hemoglobin, a slow breaking of carboxyhemoglobin compound and the prolonged elimination thereafter.

The nerve cells are most sensitive to the lack of oxygen and it is easy to understand that the first subjective and objective discomforts are noticed as disturbances in the nervous system.

Investigations carried out among young and healthy drivers in the United States showed that during one-hour exposures to carbon monoxide of about 50 ppm (equal to 58 mg/m^3 of air) their reception and observation abilities were decreased and their ability to react needed more time than usual [1]. Similar unfavorable results were obtained in investigations carried out on experimental animals. Changes in metabolism of porphyrin as well as morphological changes in isolated nerve cells were found when these animals were exposed to carbon monoxide concentration of 2 ppm for 10 weeks [1].

Similar results were obtained in the Federal Republic of Germany where exceptional attention has lately been paid to the problem of air pollution [2].

It is important to study the acute effect of such air pollutants as carbon monoxide and nitrogen oxides, in order to detect early symptoms of reduced psychophysical ability of drivers in intensive urban traffic. High concentrations of harmful gases, primarily carbon monoxide, influence the most subtle human reactions, like perception, subcortical formation of response and even the complete length of time between the perception and reaction, i.e. functions that are of extreme importance to safe driving.

2   Objective of Investigation

Studying this problem we had in mind that Yugoslavia, according to the number of traffic accidents and the number of injured persons per number of motor vehicles and number of kilometers covered, is in one of the most jeopardized places in Europe. Epidemiologic-statistical investigations in our country point out that the greatest number of accidents happen in urban areas and especially in towns that are well-known for air pollution. Our investigations

simultaneously warn us of the constant increase in carbon monoxide and nitrogen oxides concentration on these locations.

The following fields were covered in our investigation:

a) Concentration level of air pollutants originating from motor vehicles
with special emphasis on the concentration of carbon monoxide and nitrogen
oxides.

b) Concentration of carbon monoxide inside the cabin of a motor vehicle,
with a special analysis of the correlation between the concentration of carbon
monoxide outside and inside the vehicle.

3   Methodology

Air samples were taken at five selected crossroads and along the profile
of traffic arteries at the height of 1.5 meters. The crossroads were selected such
as to represent a certain characteristic of streetwidth and ascent frequency
and type of traffic as well as a certain mezzoclimate case. Simultaneously with
the sampling, an investigation of traffic frequency was carried out: mezzoclimatic characteristics were measured at the time of sampling.

4   Results and Discussion

The results obtained showed that the quantities of air pollutants found
are extremely high (Table) and that they had a special influence on the psychophysical ability of drivers; the level of concentration does not depend solely on
traffic frequency, but results from the reciprocal influence of various factors, as
gabarite streets, traffic regulation and direction, width of the street, height of
buildings, direction of the wind blowing, etc.

The measurement of carbon monoxide dispersion is shown in Figure 1.
Note that despite irregular behavior of this air pollutant we still can see certain
regularities. The pollution level, starting from the initial point, grows until on
the pavement, near the building itself, it reaches values two to three times
higher than the initial Level. It is interesting to note that the greatest concentrations were found at the height of the *third floor* which is most probably due to
certain turbulences of air masses.

We found that the pollution on the pavements is 2–3 times higher than in
the center of the square, and that the results for cross walks and the center of
the square are nearly the same. All the results should be considered in reciprocal
dependability on mezzoclimatic conditions (especially wind), on the number
and type of moving vehicles, their speed, etc. It is important to note that the
air currents made by moving vehicles push the air pollutants towards the
pavement, and in the absence of good ventilation or dispersion, and with high
buildings on both sides, air pollutants accumulate and often reach a considerable
height over ground.

Investigations of traffic policemen showed that CO concentrations in the blood of nonsmokers were ranging from 0.47 to 1.2% when CO was found in the amount of 30–40 mg/m³ in a Belgrade street while the concentrations in the blood of smokers ranged from 0.94 to 2.4%. Carboxyhemoglobin amounted to 3.8–6.2% for nonsmokers, and to 5.6–8.2% for smokers. Methemoglobin ranged from 1.9 to 2.7%.

Comparing our results with recent measurements in other European cities. Belgrade's situation must be regarded as rather serious: In Frankfurt (West Germany) the carboxyhemoglobin concentration of a group of 138 municipal street cleaners (smokers and nonsmokers) averaged only 4% [5]. Relating our data to the conclusions of Gilgen of the Swiss Institute of Hygiene that 2–8% carboxyhemoglobin concentration affects the nervous system already, the amount measured in Belgrade may indeed influence the psychophysical capabilities substantially [6].

The harmful effect of carbon monoxide represents a problem not only to inhabitants of buildings in big traffic arteries or pedestrians and traffic policemen, but it is of special importance to such traffic participants as bus or taxi drivers. In order to investigate the inside of vehicles polluted by harmful gases from the outside, we made the following experiments: A car with a driver and researcher was taken out of the city where it was ventilated for half an hour. Then the motor was started to work for 10 minutes, its gear being in the neutral position, and car set off towards Belgrade along more or less busy roads. After entering Belgrade, the car was driven into a busy street, then into a more busy one, and at last into the busiest street (Trg Republike). During the whole trip samples of air were taken and analyzed for CO. It can be seen from Figure 2 that the concentrations of this pollutant slowly grow to exceed even the maximum allowed values for working conditions in the more and the most busy streets

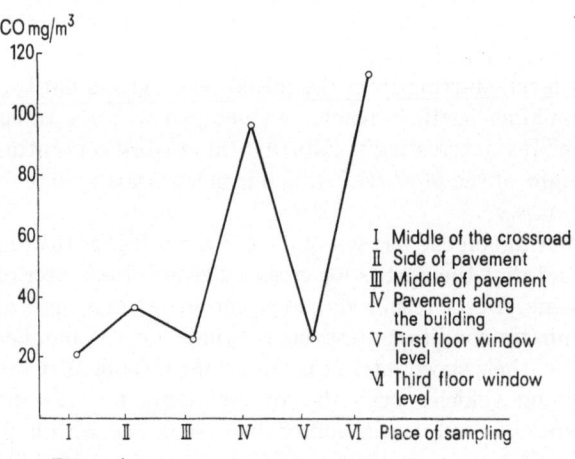

Figure 1

Carbon monoxide concentrations measured in 'London square', Belgrade (September).

in Belgrade. Under 'maximum allowed values for working conditions' we understand those concentrations of pollutants which after 8 hours of inhaling at work would not produce harmful effects to the human organism.

An investigation of a group of 10 bus drivers showed a similar increase in CO concentrations, in carboxyhemoglobin and methemoglobin alike which warned us about considerably decreased percentage of free hemoglobin for combining with O_2 and consequently about effects on the psychophysical system. Frequent complaints of drivers about air being too much polluted in the newly-opened Terazije tunnel led us to undertake a thorough investigation of carbon monoxide concentration outside and inside the tunnel. Carbon monoxide values ranging about 40 mg/m³ and up to 90 mg/m³ inside the cars leaving the tunnel which would correspond to an an average short time CO-hemoglobin concentration of 5.5% and 12.4%, respectively (using the relation 1 ppm CO → 0.16% CO-hemoglobin, as suggested by Rogers, and without accounting for accumulative processes!), warned us about an exceptional danger to drivers and their psychophysical capability especially when the traffic would stop inside the tunnel during rush hours, causing possible danger of acute poisoning.

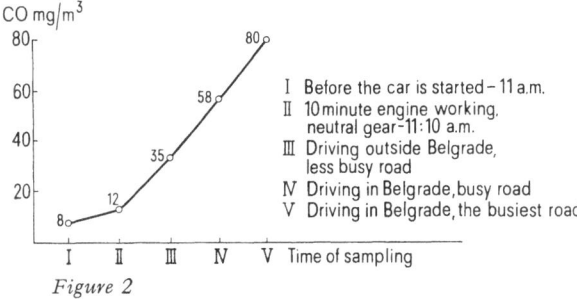

Figure 2
Carbon monoxide concentration *inside* a test car.

5   Conclusion

Analysing the data obtained in our investigations and comparing them with data from literature we see that the pollution values obtained in Belgrade are as critical as those found by Reed and Trott in selected English towns (London, Birmingham, Manchester, Glasgow) or by Pitts in Los Angeles [4].

The results of our investigations lead us to the conclusion that pollutants originating from motor vehicles might have a far greater impact on traffic accidents than generally assumed, particularly through the influence on the psychophysical capability of drivers.

In order to cope with these medical problems in an effective way, we need the cooperation of other fields of science: The combustion process and the application of filters should be improved urgently: the results we obtained showed great concentrations of carbon monoxide inside the car. The design of new streets should take the new knowledge in the field of pollution dispersion into account: Our investigation showed the accumulation of pollutants in narrow streets. Traffic planning, especially one way traffic and synchronized green lights, will help to increase the traffic flow and thus reduce the average emission per car.

All measures for protection should be considered from the broadest social aspects. These measures are neither simple nor easy to apply. They require actions with wide social impact, considerable material resources and team-work of experts in various fields. It is particularly important to exchange experience and research data on a broad international level.

References

[1] E. BLONGUIST, 3rd Air Pollution National Conference, Washington, USA (1966).
[2] H. MAY, F. J. DREYHAUPT, E. PLASSMANN and O. J. ZUNDORF, *Research on Automobile Emission in Industrial Centers*, Second International Clean Air Congress of the International Union of Pollution Prevention Association, Washington, USA (1970).
[3] L. E. REED and P. E. TROTT, *Continuous Measurement of Carbon Monoxide in Streets 1967-9, Atmospheric Environment*, Int. J. *5*, No. 1 (January 1971).
[4] J. N. PITTS, *Environmental Appraisal; Oxidants, Hydrocarbons and Oxides of Nitrogen.* J. Air Pollut. Control Ass. *19* (1969).
[5] W. BARTSCH, *Umweltschutz – Menschenschutz* (Fischer Verlag, Frankfurt 1972), p. 77, 78.
[6] W. GILGEN, in: *Schutz unseres Lebensraumes* (ETH-Symposium, Zürich 1970).

Pollution factors[1]) measured in Belgrade.

Sampling site	Square	CO	NO_x	Number of vehicles per hour
1	'Trg Republike'	12.0	2.0	2,756
2	'London'	13.0	6.8	2,932
3	'Odeon'	7.7	4.7	1,329
4	'Pošta'	9.8	12.4	3,114
5	'Slavija'	4.5	5.2	2,258

[1]) Pollution factor is a relation between concentrations found and maximum tolerable concentrations. (Present Air Pollution Law of Yugoslavia: for CO: 3.0 mg/m³, for NO_x: 0.085 mg/m³.)

Photochemical Aspects of Air Pollution in Europe

by S. J. Formosinho and A. C. Cardoso
Instituto Geofisico, Universidade de Coimbra, Coimbra, Portugal

Abstract

Todays theories on photochemical smog are discussed and new explications of the rapid oxidation of NO to NO_2 presented. It is shown that oxygen singlet state can be produced very efficiently in the quenching reactions of excited singlet and triplet states of aromatic hydrocarbons by oxygen. The importance of such processes to the chemistry of polluted atmospheres is discussed particularly with regard to the symptoms of photochemical pollution in Europe.

1   Introduction: Atmospheric Behaviour

Substances emitted to the atmosphere are subjected to a variety of physical and chemical influences. In principle, *inorganic* pollutants are light gases that can be carried away very far: A good example is the SO_2 pollution in Sweden which is due mainly to industrial activity in Great Britain and in Central Europe. (The long-range transport of pollutants is discussed in the chapter of D. Whelpdale and R. Shaw). Owing to that, the inorganic pollutants are always found in the atmosphere in much smaller quantities than the amounts calculated from the emission of known sources, and tend to participate in the small composition changes of the terrestrial atmosphere. This is not so much the case for *organic* pollutants that are not so easily transported and therefore tend to accumulate in more confined areas. The chemistry of the upper atmosphere is dominated by photochemical reactions, i.e. chemical processes due to the absorption of light by molecules and atoms. By absorption of a photon of light the chemical species acquire a different reactivity from their usual behaviour under thermal conditions. They can dissociate, internally rearrange, suffer conversion between different electronic states or they can loose energy through collisions and emission of light. Only the deactivation processes are ineffective for the initiation of some chemical step of the molecule in a primary process. The primary processes can lead directly to stable molecules or they may yield other unstable products which undergo further reaction in secondary processes. In the lower parts of the atmosphere the photochemical reactions are not so domiant due to the filter effect of ozone that limits the solar radiation reaching the ground to wavelenghts longer than 290 nm. In these low layers most inorganic gases are converted to stable salts like sulphates and nitrates which eventually lead to the soil; on the other hand the organic compounds will end up in carbon dioxide, water and, also with some compounds, in ammonium salts and nitrates. However, these chemical reactions are not instantaneous and the studies of the pollution in the air more concerned with the chain of reactions that precedes the formation of the stable compounds.

Broadly we can distinguish two basic types of pollution: the sulfur dioxide – sulfuric acid – sulfate pollution and the oxidation – type pollution. The first is the most common type in England and Central and Eastern Europe. The SO_2 emitted into the atmosphere, due to the combustion of sulfides salts and organic molecules present in oil or coal, is oxidized to SO_3 in reactions catalyzed by ash and more rapidly through some photochemical reaction. Then SO_3 reacts with water vapour and ends up in aerosols containing sulfuric acid. The second type of pollution also contains large amounts of nitrogen oxides and various hydrocarbons released by auto exhaust and other sources. In such atmospheres under strong irradiation a chain of very rapid photochemical reactions occurs with the formation of ozone, nitrogen dioxide and oxidized forms of organic compounds, particularly aldehydes. The correlation of the oxidizing power, eye irritation and plant damage of these atmospheres with the 24-hr integrated solar irradiation leaves no doubt of the light effect on the complex chemical reactions in this type of pollution.

The chemistry of the photochemical smog has been reviewed quite extensively by several authours [1–4]. It is not our intention to go into all the details, but we will try to give a general view of the kind of problems that arise in this type of polluted atmospheres. During the daylight hours, the nitric oxide is rapidly converted to nitric dioxide. The ozone content arises from levels 1–3 pphm in early morning to 20 pphm at noon. The hydrocarbon level decreases and the oxidized products, mainly aldehydes such as acrolein and formaldehyde and some peroxyacyl intrates (PAN), increase. It is the oxidized products like acrolein, formaldehyde and the PAN compounds mixed with ozone and nitrogen oxides that are mainly responsible for the eye irritant and plant damaging properties of the photochemical smog. The solar light energy is used very effectively in making free radicals that can undergo a great number of fast reactions in chain producing all the noxious products of the photochemical smog. Later these products are scavenged from the atmosphere by other chemical reactions, rainfall, etc. and eventually will be degradated in CO_2 and in H_2O, substances relatively inert to solar radiation at the ground and consequently not involved in the primary photochemical processes, although they may play a role in secondary reactions. However, the scavenging processes are much slower than the photochemical reactions and therefore they cannot prevent the build-up of pollutants in the atmosphere.

Although the atmosphere has a very high proportion of oxygen, the majority of the chemical reactions with O_2 is very slow. So what are the oxidant species in an oxidizing atmosphere? How are they formed? What are the chemical reactions and the products? Can we predict the chemical evolution of a polluted atmosphere under solar irradiation? These are some of the questions that may be asked about the photochemical smog. Presently, through the extensive studies on the photochemical smog in Los Angeles, we know reasonably well the products formed, the reactions of the inorganic compounds and some of the reactions of the organic hydrocarbons. However hypothesis over other reactions are still very speculative. The problem of the oxidizing species in a polluted atmosphere is still a matter of some controversy. Very little is known about the mechanism of aerosol formation and its importance in a photochemical smog and predictions on the evolution of the polluted atmospheres are difficult owing to the lack of data on product yields, but good forecasts can be made, for example, about levels of eye irritation in Los Angeles.

The polluted atmospheres in Europe are basically of the reducing-type but the levels of organic compounds and oxides of nitrogen are rising in Central and East Europe owing to the continuous increase in car traffic and industries. Aromatic hydrocarbons are found in every major European town, such as London, Paris, Rome, Vienna, Leningrad, Berlin, Rotterdam, Prague, Budapest, Kiev, Liege and many others [5, 6]. Recent determinations [7] on the volatile organic substances with 6–20 carbon atoms in Zurich revealed, for 22 out of the 108 molecules identified, values around 0.25 ppm from which 0.17 ppm are of aromatic compounds. Analysis [8] in Hamburg of lettuce, soybean, rye and tobacco plants grown in fields or in greenhouses showed 1,2- and 4,5-benzopyrenes, perylene, anthracene, 1,12-benzoperylene, 1,2,5,6-dibenzoanthracene and coronene, all carcerinogenic polycyclic hydrocarbons. These residues are attributed to retention of the compounds by the plants due to air pollution, since plants of the same seeds grown in special rooms in which air was supplied through filters showed none of those aromatic compounds and there was no evidence of biosynthesis of the organic compounds in the experimental conditions. The retention values went up to 4 µg/kg. Other aromatic hydrocarbons of higher vapour pressure like benzene, naphthalene, anthracene and derivatives are also present in the air not only in aerosols but as vapours. In spite of lacking favourable metereological conditions, particularly of a strong solar irradiation, with such levels of hydrocarbon pollution and also of nitrogen oxides coming from the same sources, it is not surprising that some effects characteristic of polluted oxidizing atmospheres are already occurring in Europe. The formaldehyde level in Berlin [9] averages to 13.4 µg/m³ in summer, whereas in winter the levels averaged only 2.7 µg/m³. The PAN has been identified in the Gulf of Genoa and in the Netherlands [10]. Frankfurt already shows some photochemical smog pollution [11], although much milder than in Los Angeles, and analyses for oxidant samples in Moscow and elsewhere in USSR give indications of oxidant in the air and of photochemical smog as a potential problem [12].

Not much is known about the effect of aerosols on photochemical reactions, but in relation with the pollution condition in Central Europe, some research is already being done in Bonn [13] on these effects. It has already been shown that metallic surfaces can photosensitize photochemical reactions with visible light of long wavelengths. So, although in Central Europe the photochemical smog is not yet a problem despite the continous increase of car traffic and industrial pollution, some symptoms of the evolution of the reducing polluted European atmospheres into polluted oxidizing areas have already been detected. In thenorth of Europe, particularly in Norway and Finland, the hydrocarbon pollution is still low.

The air pollution problem both regional and local is in general much more serious in Central and Northern Europe than in the Mediterranean areas and in the south. However, during special weather conditions in high pressure cells the concentration of pollutants may increase considerably even in the southern regions. These conditions associated with strong solar irradiation and hot weather are very favourable to the formation of an oxidizing polluted atmospheric environment. Such conditions seem to exist for example in Lisbon [14] where the sulphur dioxide pollution is associated with overhead inversions, mostly in spring and summer, contrary to what is normally observed in other European towns, but similar to the condition in Los Angeles. The pollution in Lisbon is mainly from industry and car traffic but is still less considerable than in the main towns in Central Europe. However, some hydrocarbons have already been detected [15] and nitrogen oxides must exist in reasonably high levels on certain days [14]. Sources of hydrocarbons which could be important in relation to other emissions, particularly in Southern Europe, are the highly aromatic bushes and trees. Went [16] estimated the total emission of terpene-like and other hydrocarbons by several kinds of forests in the USA to be as high as 17×10^7 ton/year. Non-urban regions in California with a high density of trees such as pines or citrus fruit show this kind of photochemical smog due probably to energy transfer from photo-excited hydrocarbons to oxygen, followed by the formation of ozone and subsequent oxidation of the hydrocarbon by the ozone.

It can be said that photochemical smog is not yet an acute problem in Europe. The question which should be asked, however, is if we will see some strong oxidizing polluted atmospheres in Europe in a couple of years? For this sort of question it is necessary to know a lot of details about the chemistry of the photochemical smog. One of the fundamental problems is a clear elucidation of the nature of the oxidants in these atmospheres and how such species can be formed. Some research that we have done on the photochemistry of aromatic molecules in the vapour phase seems to be relevant to this problem. This will be the subject of the discussion in the second part of this paper.

2   Energy Transfer As a Source of Singlet Oxygen

Although considerable understanding of the photochemical processes in polluted atmosphere has been attained [2, 3] there are several crucial aspects of the problem not yet fully explained. One of the most important questions is the rapid oxidation of NO to NO_2 with the concurrent disappearance of the hydrocarbons and the build-up of oxidants, namely of ozone, aldehydes and peroxyacyl nitrates. The reaction,

$$2 \, NO + O_2 \rightarrow 2 \, NO_2$$

is not fast enough to account for the build-up of NO_2. The way in which ozone does not appear until almost all the nitrogen monoxide has been consumed, suggests strongly that the relatively rapid process

$$NO + O_3 \rightarrow NO_2 + O_2$$

is the major loss of ozone when NO is present. However, since O_3 is formed mainly through the reaction

$$O + O_2 + M \rightarrow O_3 + M$$

and atomic oxygen, in the lower layers of the atmosphere, has been supposed to originate mainly from the photolysis of O_3, we will be left with the unsolved problem of explaining the source of ozone.

Recently it has been recognized that electronic excited singlet oxygen $^1O_2$ (probably in the $^1\Delta_g$ state) could play an important role in the oxidation of NO to NO_2. The mechanism suggested by PITTS et al. [17] involves the oxidation of olefins by $^1O_2$ forming peroxide and subsequently radicals

$$^1O_2 + -\underset{\underset{H}{|}}{C}=C-C- \rightarrow -C-\underset{\underset{OOH}{|}}{C}=C \xrightarrow{\text{thermal}} \text{radicals (e.g., RCO·)}$$

due to thermal decomposition of those unstable compounds. Once such radicals are formed they can oxidize NO through several reactions [2] such as

$$RCO· + O_2 \rightarrow RCO_3· \quad RCO_3· + NO \rightarrow RCO_2· + NO_2, \tag{i}$$

$$RCO_3· + NO_2 \rightarrow RCO_3 \, NO_2. \tag{ii}$$

Process (ii), much less frequent than (i), leads to PAN. NO_2 can be photolysed by light of wavelength shorter than 400 nm

$$NO_2 + h\nu \rightarrow NO + O$$

and once O is formed ozone could rapidly build up.

The possibility that singlet molecular oxygen was involved in the oxidation of NO was first discussed by LEIGHTON [1], but remained a problem until some reasonable mechanism for the formation of $^1O_2$ could be thought of. There are several possible mechanisms that have been discussed by several authors [17, 18]. One of the most important seems to be the energy transfer process where $^1O_2$ is formed on collisions of O_2 molecules with an excited sensitizer. Such a mechanism is usually depicted

$$S_0 + h\nu \rightarrow S_1 \tag{1}$$

$$S_1 \xrightarrow[\text{crossing}]{\text{intersystem}} T_1 \tag{2}$$

$$T_1 + O_2(^3\Sigma_g^-) \rightarrow S_0 + O_2(^1\Delta_g \text{ or } ^1\Sigma_g^+) \tag{3}$$

$$\text{or } [S_0 + O_2(^3\Sigma_g^-)] \tag{4}$$

where S_0 is the ground electronic state of the sensitizer and S_1 and T_1 are the lowest excited singlet and triplet states. $^3\Sigma_g^-$ ist he ground state of the oxygen molecule and $^1\Delta_g$ and $^1\Delta_g^+$ the first and second electronic excited singlets above the ground state, respectively. Process (3) can only occur when the difference in energy of T_1 and S_0 is higher than the electronic energy of $^1O_2(^1\Delta_g$ 7.882 cm^{-1}(1,269 nm) and $^1\Sigma_g^+$ 13.120 cm^{-1} (762 nm)].

The sensitizer molecules must absorb at wavelengths longer than 290 nm and the most important ones in the polluted atmosphere are the aromatic hydrocarbons which have very high extinction coefficients ($\varepsilon \simeq 10^4$ mol^{-1} l cm^{-1}) and, with less importance some carbonyl compounds. Through such a mechanism the quantum yield of formation of singlet oxygen (number of $^1O_2$ formed per photon absorbed) is $\phi\,^1O_2 < \phi_{T_1}$ where ϕ_{T_1} is the quantum yield of triplet formation. However, we have some experimental evidence that $^1O_2$ can be also produced in the vapour phase in the quenching process of excited singlet states.

The excited singlet state S_1 is quenched by oxygen very efficiently; in the case of the aromatic molecules the quenching is diffusion controlled. The most probably processes of this quenching are

$$S_1 + O_2(^3\Sigma_g^-) \rightarrow T_n + O_2(^1\Delta_g \text{ or } ^1\Sigma_g^+) \tag{5}$$

$$S_1 + O_2(^3\Sigma_g^-) \rightarrow S_0 + O_2(^1\Delta_g \text{ or } ^1\Sigma_g^+) \tag{6}$$

$$S_1 + O_2(^3\Sigma_g^-) \rightarrow T_n + O_2(^3\Sigma_g^-) \tag{7}$$

$$S_1 + O_2(^3\Sigma_g^-) \rightarrow S_0 + O_2(^3\Sigma_g^-) \tag{8}$$

where T_n is an excited triplet state of the sensitizer, not necessarily the lowest state T_1. Processes (5) and (6) are energy transfer processes where $^1O_2$ can be formed. Process (6) is always energetically possible, but (5) can only occur when the difference in energy between S_1 and T_n is greater than the energy of any oxygen singlet state.

The emission of O_2 ($^1\Delta_g$) has been observed by several authors on irradiations of mixtures of organic vapours and oxygen. SNELLING [19] has investigated the problem with benzene; KUMMLER and BORTNER [20] with benzaldehyde and STEER et al. [21] with naphthalene. KEARNS et al. [22] and WASSERMAN et al. [23] have also observed O_2 ($^1\Delta_g$) by electron spin resonance with low pressures ($\simeq 1$ Torr) gaseous solutions of naphthalene, quinoxaline and perfluoronaphtalene in oxygen.

KEARNS and co-workers [22] have detected O_2 ($^1\Delta_g$) when naphthalene was irradiated with light from a mercury lamp at room temperature in presence of 0.1 to 0.3 Torr of O_2. The quantum yield of O_2 ($^1\Delta_g$) was found to be 0.5 ± 0.3. Those workers have attributed the O_2 ($^1\Delta_g$) formation to reaction (3). However in these low pressures conditions, it has been shown by ASHPOLE, FORMOSINHO and PORTER [24, 25] that no appreciable naphthalene triplet T_1 ($\phi_{T_1} < 0.05$) is formed on irradiation. Therefore the formation of O_2 ($^1\Delta_g$) can only be due to the quenching of the excited singlet state S_1 or to the quenching of an upper triplet state. This last process is very inefficient since such a triplet state is very short lived ($\tau \simeq 2$ ns) [26, 27] and the quenching of the triplet state by oxygen in the vapour phase is about $^1/_{10}$ to $^1/_{100}$ of the collision controlled rate [28, 29]. In fact, in those low pressure conditions, FORMOSINHO and PORTER [26, 27] have shown that the intersystem crossing process is completely reversible and the only possible source of O_2 ($^1\Delta_g$) are processes (5) and (or) (6).

MORIKAWA and CVETANOVIĆ [29] have found that at overall pressures of 20 Torr, oxygen quenches benzene fluorescence without noticeable change in the concentration of the lowest triplet, i.e., the yield of singlet quenching via process (5) is almost equal to the triplet yield of the unperturbed molecule. The same has been observed in anthracene [26, 30] and naphthalene [26, 27] for the total yield of triplets formed for pressures between 0.5 Torr and 20 Torr. In the cases of naphthalene and anthracene (26, 27, 30) process (5) seems not to be a very efficient source of O_2 ($^1\Delta_g$). So, probably quenching of the excited aromatic first excited singlet state occurs mainly via process (6) and (7) with a yield of $^1O_2$ ca. 1-ϕ_{T_1} (ϕ_{T_1} triplet yield). In fact, the $^1O_2$ ($^1\Delta_g$) yield determined by KEARNS et al. [22] agrees with the fluorescence and the internal conversion yields of naphthalene (0.45) for an average excitation at 250 nm where the emission of the mercury lamps is strongest. The same conclusion seems to be true at other wavelengths of excitation up to 310 nm [26, 27]. In the studies [19] of O_2 ($^1\Delta_g$) emission with benzene and oxygen, the emission increased up to O_2 pressures of 1 Torr, and there was no further increase for higher pressures. Since the fluorescence of benzene is only 30% quenched at 1 Torr, SNELLING concluded that $^1\Delta_g$ was produced by quenching of triplet benzene rather than singlet benzene. However, in view of the previous discussion and others results unknown when SNELLING had his work published, that assumption must probably need some revision. In fact, since it is known that oxygen quenches the fluorescence of benzene without changing the concentration of the lowest triplet for pressures around 20 Torr, if the $^1\Delta_g$ state was only produced via process (3) the emission should have attained its maximum for much lower oxygen pressures ($\simeq 0.1$ Torr) since the triplet lifetime is 10^{-3}s [31] and the oxygen quenching rate for the benzene triplet is 1.26×10^{10} mol^{-1} s^{-1} [29]. The triplet lifetime of benzene obtained by SNELLING through his data is 2.6×10^{-5}s, a value two orders of magnitude shorter than the value of BURTON and HUNZIKER, which is in good agreement with the triplet lifetimes of unimolecular decay of other aromatic molecules [26, 32, 33]. However, the singlet lifetime of benzene is too short to explain the rapid increase in $^1\Delta_g$ emission from 0.01 to 1 Torr of O_2 in terms of singlet formation only by quenching of the benzene singlet. So it seems that in the benzene case, and probably with the other aromatic molecules too, $^1O_2$ is formed in part in the oxygen quenching of the singlet state and also in the quenching of the lowest triplet state. This is supported also by the studies of SNELLING with pressures of 10 Torr of benzene. The results are similar to the ones corresponding to 20 Torr but are moved to lower oxygen pressures in agreement with the increase of the fluorescence lifetime of benzene for pressures below 20 Torr [34]. In any case, a slight increase ($\simeq 0.4$) in $^1\Delta_g$ emission should have been observed in both cases up to pressures of 10 Torr of O_2 where the benzene singlet is completely quenched. However, a small rise in O_2 ($^1\Delta_g$) production at high O_2 pressures might be compensated for by a decrease by some quenching process at relatively high O_2 ($^1\Delta_g$) concentrations.

One of these processes could be the almost diffusion controlled quenching of O_2 ($^1\Delta_g$) by excited triplet and probably singlet state of aromatic molecules as was shown by DUNCAN and KEARNS [35]. However, their data, at least for naphtalene, owing to the low pressures used, should imply a quenching mainly by a high triplet state T_n.

O_2 ($^1\Delta_g$) state is a relatively long lived species, quite stable to collisional quenching by molecules in their ground states [36, 37]. However, the $^1\Sigma_g^+$ state is rapidly quenched to $^1\Delta_g$ and because of that, has eluded direct detection. Nevertheless, ANDREWS and ABRAHAMSON [38] have recently shown the formation of O_2 ($^1\Sigma_g^+$) in the quenching of fluoronaphthalene by oxygen. For energetic reasons, process (5) is excluded as a possible source, but, since for the lowest oxygen pressures (13 Torr) used the singlet of benzene is completely quenched, it is not possible to decide between (3) and (6); there is probably some contribution by both processes.

In conclusion, O_2 ($^1\Delta_g$) seems to be formed by an energy transfer mechanism in the oxygen quenching of excited singlet and triplet states of aromatic molecules. The quantum yield of formation by such a mechanism is high (ca. 1.0) and almost independent of the yield of triplet formation. This evidence makes the estimation of PITTS et al. [17] on the upper limit (80 pphm/hr) of $^1O_2$ formation in the atmosphere of Los Angeles more reliable, since they have considered a quantum yield of 1.0 for O_2 ($^1\Delta_g$) production by energy transfer from excited aromatic molecules.

3   Conclusion

The elucidation of the processes that prevail in the transformation of a typical polluted reducing atmosphere into an oxydizing region is an important step to predict the evolution of the polluted atmospheres in urban regions. This work tries to show the importance of the energy transfer mechanism of aromatic hydrocarbons to oxygen in the production of exicted singlet oxygen. This will be the primary oxidant species in the photochemical smog, oxidizing relatively rapidly other organic and inorganic compounds. Since the directly photoexcitation of O_2 to $^1O_2$ is a very inefficient process, once there is some sensitizer which can absorb light at wavelengths longer than 290 nm, the necessary conditions and ingredients to start a photochemical smog exist even in the absence of nitrogen oxides. The importance of aromatic compounds in this process depends on their concentration in any particular area. Other sensitizers certainly exist. One possibility are the organic metallic compounds present in aerosols of certain industrial areas. These compounds absorb at much longer wavelengths than the aromatic hydrocarbons and although they do not absorb very strongly in these regions of the light spectrum, such a factor could be compensated for in part by a more intense solar irradiation on earth at these wavelengths. However, not much is known about this topic. Obviously, we have only dealt with a very small part of the complex chemical problems of photochemical smog since so much is still unknown.

The symptoms in Europe are all pointing towards an increase in the kind of photochemical pollution that already mildly exists. If the energy transfer mechanism is really as important as it seems to be, such a prediction is obvious since the amount of hydrocarbon pollutants is increasing every year. Naturally, the kind of oxidizing atmosphere, which will appear in Europe, will not be exactly the same as in Los Angeles, due to the differences in metereological conditions. However, some form of photochemical smog is bound to appear during summertime associated with the reducing-type pollution, which becomes stronger in the winter season. In fact, a similar problem exists in Los Angeles since the type of pollution that prevails in the winter is distinct from the one in summer and fall [39].

A great deal more needs to be done to understand the physical and chemical processes in the atmosphere. In such research a closer association between physicists, chemists and metereologists is very necessary. The medical aspects of pollution are certainly the field where the most research is being done. Much more will be done, surely, but we would very much welcome more research associated with the chemical problems of pollution, particularly concerning new kinds of pollutants dangerous to life, which can emerge from the continuous evolution of the polluted atmospheres.

Oxidizing atmospheres are very complex, having a great number of different sources of pollutants not so easily dispersed in the air in adverse metereological conditions. Therefore it is not surprizing that the photochemical smog is more difficult to control than a SO_2 pollution, and this seems to us a point that should be taken into consideration in pollution prevention.

Acknowledgments

This work is included in the research project CQ-2 of the CEQNR (Chemical Laboratory, University of Coimbra) and was done in collaboration with Serviço Meteorológico Nacional. We tank the director of SMN, Dr. A. Silva de Sousa, for his interest and support. Thanks are also due to the director of Instituto Geofisico and CEQNR, Prof. Dr. F. Pinto Coelho, for his interest and for many helpful discussions.

References

[1] P. A. LEIGHTON *Photochemistry of Air Pollution* (Academic Press, New York 1961).

[2] A. P. ALTSHULLER and J. J. BUFALINI, Photochem. Photobiol. *4*, 97 (1965).

[3] A. P. ALTSHULLER and J. J. BUFALINI, Envir. Sci. technol. *5*, 39 (1971).

[4] R. D. CADLE and E. R. ALLEN, Science *167*, 243 (1970).

[5] A. P. ALTSHULLER, Anal. Chem. *41*, 1R (1969).

[6] A. P. ALTSHULLER, Anal. Chem. *39* 10 R (1967).

[7] K. GROB and G. GROB, J. Chromatogr. *62*, 1 (1971).

[8] G. GRIMMER and G. DUVEL, Z. Naturf. *25b*, 1171 (1970).

[9] E. LAHMANN and K. JANDER, Gesundh. Ingr. (Munich) *89*, 18 (1968).

[10] P. CHOVIN and A. ROUSSEL, *La Pollution Atmosphérique*, (Presses Universitaires de France, Paris 1968).

[11] K. H. BECKER, Chemie unserer Zeit *5*, 9 (1971).

[12] V. A. POPOV, Hyg. Sanit. *31*, 3 (1966).

[13] D. BREUER and H. MOESTA, quoted in [11].

[14] A. G. EJARQUE and M. DE QUINTANILHA, Rev. Port. Quim. *6*, 125 (1964).

[15] E. V. ALBERTO, *6*, 78 (1964).

[16] F. W. WENT, Proc. nat. Acad. Sci. USA *46*, 212 (1960).

[17] J. N. PITTS, Jr., A. U. KHAN, E. B. SMITH and R. P. WAYNE, Envir. Sci. technol. *3*, 241 (1969).

[18] R. H. KUMMLER and M. H. BORTNER, J. geophys. Res. *75*, 3115 (1970).

[19] D. R. SNELLING, Chem. Phys. Lett. *2*, 346 (1968).

[20] R. H. KUMMLER and M. H. BORTNER, Envir. Sci. technol. *3*, 944 (1969).

[21] R. P. STEER, J. L. SPRUNG and J. N. PITTS, Jr., Envir. Sci. technol *3*, 946 (1969).

[22] D. R. KEARNS, A. U. KHAN, C. K. DUNCAN and A. H. MAKI, J. Am. chem. Soc. *91*, 1039 (1969).

[23] E. WASSERMAN, V. J. KUCK, W. M. DELEVAN and W. A. YAGER, J. Am. chem. Soc. *91*, 1040 (1969)

[24] C. W. ASHPOLE, S. J. FORMOSINHO and G. PORTER, Chem. Comm., p. 1305 (1969).

[25] C. W. ASHPOLE, S. J. FORMOSINHO and G. PORTER, Proc. R. Soc. Lond. *A 323*, 11 (1971).

[26] S. J. FORMOSINHO, Ph. D. thesis (London 1971).

[27] S. J. FORMOSINHO and G. PORTER, Proc. R. Soc. Lond. [A], in publication.

[28] G. PORTER and P. WEST, Proc. R. Soc. Lond. [A] 279, 302 (1964).

[29] A. MORIKAWA and R. J. CVETANOVIĆ, J. Chem. Phys. *52*, 3237 (1970).

[30] S. J. FORMOSINHO, G. PORTER and M. A. WEST, Proc. R. Soc. Lond. A, in publication.

[31] C. S. BURTON and H. E. HUNZIKER, Chem. Phys. Lett. *6*, 352 (1970).

[32] C. W. ASHPOLE, S. J. FORMOSINHO, G. PORTER and M. A. WEST, J.C.S. Faraday, in publication.

[33] W. H. VAN LEEUWEN, J. LANGELAAR and J. D. W. VAN VOORST, Chem. Phys. Lett. *13*, 622 (1972).

[34] M. NISHIKAWA and P. K. LUDWIG, J. Chem. Phys. *52*, 107 (1970).

[35] C. K. DUNCAN and D. R. KEARNS, Chem. Phys. Lett. *12*, 306 (1971).

[36] R. P. WAYNE, Adv. Photochem. *7*, 311 (1969).

[37] D. R. KEARNS, Chem. Rev. *71*, 395 (1971).

[38] L. J. ANDREWS and F. W. ABRAHAMSON, Chem. Phys. Lett. *10*, 113, (1971).

[39] W. J. HAMMING, W. G. MACBETH and R. L. CHASS, Arch. Envir. Health *14*, 137 (1967).

Ecoscience II:
The Pollution – Sociosphere Interface

by Jan-Olaf Willums

Very few have realized that we are not only endangering our present and future environment, but that we are also breaking the bridges to our past history and cultural heritage: treasures of irreplacable value, not measurable in economic units of any kind, are being seriously damaged or even completely destroyed by air pollution. Few persons have been so involved with these important questions as Joseph Riederer, whose outstanding research at the world famous Bavarian 'Staatsgemäldesammlung' in Munich represents an urgent message to the world to take appropriate actions to prevent the destruction of the irreplacable cultural treasures of the past.

Today's political action is greatly influenced by economic considerations. But until now economists have only in rare cases tried to include ecological aspects in a consistent way. When these aspects are applied to our economic behaviour, it is hard to avoid the conclusion that present economic trends are diametrically opposed to the requirements of the stability of our ecosystem. The answer is not to abolish growth completely and develop an immobile society, but to redirect growth and expansion: to find a strategy of sustainable growth. The traditional economic theory has so far proved inadequate for the solution of the environmental crisis. The German economist Erich Hoedl believes that the most important aspect so far excluded in economic analysis is the social component, or the 'macro-sociology'. If environmental policy is guided by economic theory as understood today, pollution will be 'repaired' rather that prevented. The shift from *economic theory* to *political economy* makes possible the integration of economic and technological aspects as well as the needed cooperation between natural and social sciences. In his article 'The New Economic Aspects of Air Pollution' he outlines his theory of a political economy able to cope with the complex problems of pollution.

The new economic theory must also account for international ecological interactions, a problem which becomes more and more important in Europe, where pollution is still regarded as a matter of national and local concern. International cooperation in this field is most urgently needed, as Andreas Uhlig of the University of Zürich points out in his article on 'Effects of Air Pollution and Air Conservation on International Economic Relations'. A simple economic model leads him to the conclusion that shortsighted national measures against polluting production processes may both influence the international flow of products and capital as well as cause a cross border transfer of polluting industries and a new heavy concentration of pollution emission, subject to the severe problems of long-range pollution transfer discussed in the first part of this book. A whole set of international legal problems thus arises.

One important step forward in the question of International Law has been the 1972 Stockholm Declaration on the Human Environment. For the first time, the international responsability of states is extended from the old territorial types of damages to environmental damage in areas beyond the limits of national jurisdiction. The legal problems of trans-frontier air pollution and the recent trends in international law are presented by PETER S. SAND, a former Associate Professor of Law at McGill and now legal officer at the FAO in Rome. Under existing legislation highly polluting industries may be tempted to relocate where regulations are not yet so strict. They can shift their capital, production and jobs to other countries to avoid pollution control costs. Such a prospect makes us wonder if 'countries of pollution convenience' might develop much like flags of convenience did in international shipping. Carried further, such consequences can affect a country's economic development and its balance of payments.

This raises immediately the question of pollution in developing countries. These countries face a different set of circumstances. With their shortage of capital they might give priority to productivity, growth and other issues of economic development rather than to what they might call 'quality-of-life luxuries'. These countries will also be suspicious of any attempt by the rich countries to coordinate pollution control policies, as they will see it as an attempt to hamper their development toward economic and political freedom. The problems faced in introducing abatement programs in developing countries are so manifold, that effective solutions seem to be doomed to fail for a long time. The technological aspects are only one of the many crucial problems, MICHAEL MCGARRY of the Asian Institute of Technology in Bangkok concludes: The social and administrative difficulties are mostly the real obstacles. The career advancement system in the civil service has little incentive for new ideas, and the Third World's political system does not encourage civic action pressure groups as is the case in the Western society. Here it becomes indeed clear how complex the fate of an ecosystem is related to man's life and behaviour. The understanding of all interactions in this multidimensional system is far from sufficient.

Pollution Damage to Works of Art

by Josef Riederer
Doerner-Institut, Bayerische Staatsgemäldesammlungen, Munich, Federal Republic of Germany

Abstract

The damaging effects of various pollutants found in European cities on different materials used for works of art call for urgent action in order to save irreplacable treasures from destruction. Today's knowledge of such destructive processes is presented in this chapter, and methods for effective protection and restoration suggested and discussed.

1   Introduction

The damage of man's environment has been the topic of much study. Although recent studies also try to account for the damage to health and well-being, which is difficult to express in terms of money, most reports totally neglect the 'ethical' aspects of the polluted environment. Many of man's greatest works of art are seriously threatened by pollution; they cannot be recreated as can, in many cases, the natural environment. Their destruction by pollution represents a permanent loss to mankind.

In this chapter, the damages to works of art by air pollution and the methods existing today for restoration and preservation are discussed. The Clean Air Commission of the Association of German Engineers (VDI), established in 1955, quite early in its activities devoted a significant portion of their efforts to research on the effects of gases and dust on materials. In 1966, the first research project on damage to works of art was conferred upon the author. It concerned the investigation of damage to outdoor bronzes due to polluiton. The project was finished in 1971 and continued with the elaboration and testing of methods for the conservation of bronzes corroded by the aggressive compounds in the air. This project will be finished in 1973 and continued with research on other metals used in works of art.

In 1972, Schimmelwitz from the Federal Institute for Testing Materials in Berlin was put in charge of the investigation of the ceday of natural stone on buildings caused by air pollution. Several other projects concerning the decay of stone and the methods for conservation are supported by German foundations.

In 1972, Frenzel, a glass restorer in Nürnberg, started to work on damages done to glass paintings with the aim of improving methods for the conservation and preservation of glass.

2   **Destructive Agents in the Air** (Air Pollutants)

During recent years, extensive material on the amount of aggressive gases in the air has been compiled. In Germany, 3.6 million tons of SO_2, 8 million tons of CO and 2 million tons of both hydro-carbons (C_nH_m) and nitrous oxides (N_xO_y) were released together with 4 million tons of dust in 1969. Of these air pollutants, only sulfur dioxide and dust are so far known to do harm to works or art.

In most of the fuels sulfur is present. 80% of the total amount of sulfur in Germany is due to the combustion of industry and the fuels used in heating houses and 20% is produced by chemical industries and foundries. Automobile exhaust is of no considerable importance for the production of sulfur dioxide.

An analysis of the dust, precipitated in Munich, showed that it contained 25% carbonized substances, 23% vitreous and amorphous compounds, 30% calcite, 20% quartz and 2% iron oxides. The action of sulfur dioxide, together with humidity and dust, causes serious damage to works of art which are exhibitet outdoors.

Figure 1
City of Vienna: Sulfur dioxide and soot from combustion are the greatest contributors to the decay of monuments and sculptures.

3   Damage to Stone

Buildings and sculptures made of natural stone are presumed to be seriously threatened by sulfur dioxide pollution which is able to transform calcite, one of the most common rock-forming materials, into gypsum which is rapidly worn down by weathering. Limestone and marble in particular, which consist entirely of calcite, and some kinds of sandstone where the grains are cemented by calcite, are subject to greater damage than ordinary sandstones which contain no minerals susceptible to aggressive water.

If we regard the extent of stone decay in Germany, we find that the 'theoretically' resistant *ordinary* sandstones, which in some regions were used to construct whole towns, are extensively destroyed, while limestone buildings, which should be easily attacked, are almost without any damage, even in the center of an extremely polluted town like Munich. It thus becomes obvious that there are still other forces causing the decay of stone, in particular natural weathering which depends on the quality of the stone and its resistance. Unfortunately many constructions such as the sculptured ornaments of churches were made of soft stones. These are quickly attacked by frost and insolation,

Figure 2
Apostle in sandstone from the Cathedral of Bamberg: Most decay on objects of stone is due to natural weathering.

making the surface brittle so that the loosened grains can be removed by the corrosive action of wind and rain. As a result, much of the really serious destruction to our cultural monuments is due more to the natural weathering of soft stones than to an attack of air polluting substances. It is the same in Venice where much of the decay is due to the weathering and the action of sea water and not to the gases from industry or motor boats. It should however be noted that pollution might have increased the destruction.

Damage to limestone primarily by chemically reactive water rich in sulfur dioxide was found only under exceptional circumstances. In southern Bavaria, a red limestone was preferred for portals and tombstones. This stone has a nodular fabric where hard nodules are separated by thin layers of clay. Aggressive water can penetrate through these layers deep into the stone where salts crystallize, pressing the nodules out to the surface. Such a considerable loss of substance leads to the rapid destruction of the stone.

The destruction of limestones which have a platy fabric occurs in much the same way. For stone constructed with plates parallel to the surface, the agressive water easily penetrates through the cleavages between the plates which burst off, one after another, by the pressure of crystallising salts.

There is no doubt that in this type of destruction, which can happen only if water is present, all the other damaging effects of water in the stone occur, especially the action of frost. Thus we see here, too, that air pollution is only a partial cause of the decay.

Figure 3
Epitaphs of red limestone on the Cathedral of Munich: Under exceptional circumstances, limestones can be attacked by water rich in sulfur dioxide.

For stone conservation, two problems have to be solved: First the brittle surface of the decayed stone has to be strengthened and secondly, the surface has to be protected so that aggressive water can no longer penetrate into the interior of the stone. Both problems can be solved with our present technica knowledge but one must realize that all the destructive forces usually continue to attack the stone, even after conversation. The result is that after some time, depending on the quality of the stone and the intensity of the destructive forces, usually after several years, the protective treatment has to be repeated.

For the strengthening of the decayed stone, synthetic materials, especially solutions of epoxy resins, silicon resins and metacrylates, have proved to be effective. In southern Germany a great number of sculptures and portals which suffered considerable damage from weathering have been protected with success. Such synthetic materials are not used for the consolidation of decaying stone on facades, for these can be protected with either organic or inorganic solutions of silicates. Among the organic silicates, the results with silicate esters have been the best.

Synthetic materials and solutions of silicates can be combined with water repellent materials which prevent aggressive water from entering the

Figure 4
Sculptures which are made of porous limestones can be attacked by agressive solutions which are produced by catalytic effects within the black crusts of soot.

stone and causing damage by dissolving the minerals, salt crystallization and cracking by frost.

In France, products have been developed which transform the sulfur dioxide of the polluted air or already formed sulfates into insoluble compounds. Furthermore, such products contain compounds which prevent the dissolution of calcite and prevent the freezing of water by lowering the freezing point to $-22\,°C$. In this way, the stone is protected from most aggressive attaks, including the action of air polluting sulfur dioxide.

More serious than the corrosive action of sulfur dioxide is damage caused by soot which settles on the surfaces of the stone. It adheres closely to the stone, penetrating deep into its pores and fissures. Only on those parts which are entirely exposed to rain is the soot washed off. On all the other parts, it remains, gradually forming thick crusts. Below these crusts the stone decays by the action of the acid compounds of the soot. The damage resulting from the various methods of removing the encrusted soot is as serious as the damage caused by the corrosive action of the soot itself. Air blasting or cleaning with acids always lead to a loss of substance of the stone.

4   Damage to Wall Paintings

In many historic towns of Germany, facades are decorated with wall paintings. Also in modern architecture, the coloration of buildings has attained importance.

Figure 5
Wall painting of the battle of Sendling by LINDENSCHMIDT (1831): A layer of soot covers the painting.

On early wall paintings, a plaster of lime was used which can be transformed by the sulfur dioxide into gypsum. This and other sulfates may destroy the paint layer or may lead to a crumbling of the surface. A destruction of pigments is rarely found since resistant pigment materials were always used. Great difficulties in the restoration of wall paintings are due to the layer of soot on their surfaces which must be very carefully removed in order to avoid harming the original painting.

At the end of the nineteenth century a special technique for painting on walls was developed, the so called silicate technique. A soluble silicate was used as a medium for the pigments. It is so resistant to weathering that there still exist a great number of wall paintings even in polluted towns from the nineteenth century which need not yet be restored.

5   Damage to Painted Glass

Painted glass windows have suffered serious damage as a result of dust and sulfur dioxide. It is supposed that a great deal of early European windows

Figure 6
Painted window from 1166 at St. Patrokli, Soest, condition in 1943 (Photo: Landeskonservator Westfalen – Lippe).

were lost, for the decay proceeded so rapidly that the few restorers able to do this difficult work had to restrict their efforts to only the most important pieces. Like the decay of stone and wall paintings, the painted windows were also attacked by natural weathering. The analysis of the corrosion products, which were sulfates, above all gypsum, showed again the destructive efficiency of the sulfur dioxide. First the black paint, a layer of fractions of a millimeter on top of the colored glass, is lost and with it all the artistic expression of the glass painting. On the glass too, damage occurs. During the first stage, the glass becomes more and more opaque, as potassium, one of the main compounds of glass, is dissolved, leaivng only silica. At the same time on the surface, a layer of 'weather-stone' is formed which consists of dust, baked together with the sulfates formed by the corrosion of the glass by the action of aggressive water. Below this layer of dust and corrosion products, little craters appear in the glass around the impurities and from these, networks of cracks spread all over the glass so that it is entirely lost.

Methods for the conservation of painted windows have been improved and developed. The black paint on the colored glass is fixed either by means of molten glass, which has a low melting point, or by synthetic resins which are

Figure 7
The same window as
Figure 6, condition in 1967
(Photo: Landeskonservator
Westfalen – Lippe).

much easier to apply. Windows which are already broken into pieces are glued together with wax or synthetic materials and then preserved between two sheets of transparent glass.

6   Damage to Bronzes

Damage of an unimaginable extent has been found on bronze sculpture. This kind of destruction was not noticed for a long time for it was covered by a thick crust of dirt. On those parts where the layer of soot is the thickest, corrosion proceeds very rapidly and leads to pitting.

The first step of the research project concerning corrosion of bronzes by air pollution was to take an inventory of sculptures exposed outdoors. In Germany there is still one bronze portal from the eighth century and two others from the eleventh century. In Braunschweig there is the great monument of a lion from 1166. More than 100 bronze sculptures from the sixteenth and seventeenth centuries are still outdoors, 53 of which are in the yards and on the facades of the residence in Munich. There are also famous fountains at Augsburg

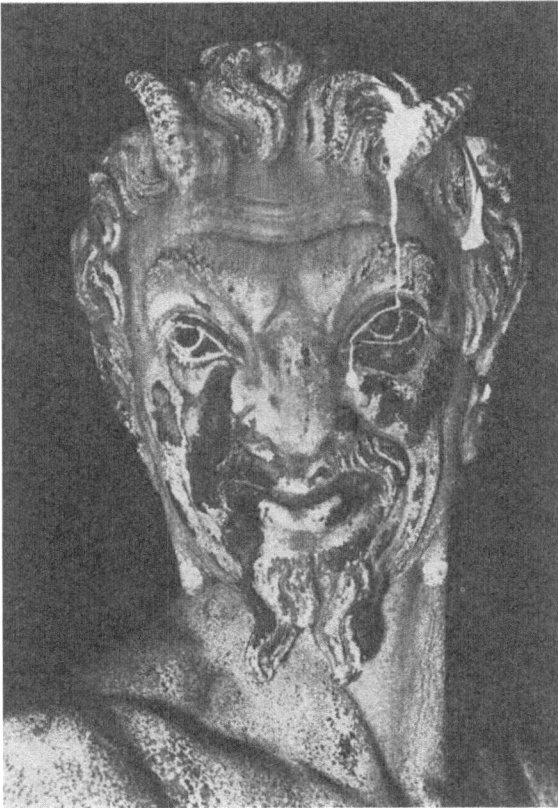

Figure 8
Satyr in bronze (1620) from the Residenz of Munich: Below the black crusts of soot the metal is transformed into sulfates.

and Nürnberg. From the eighteenth century, no more than five sculptures still exist. In the nineteenth century and the first decades of the twentieth century, a great many bronze monuments for sovereigns, generals, poets or scientists were cast and are exhibited in the yards of museums and parks. The metal was analyzed from all of the earlier bronzes up to the seventeenth century, to find out if corrosion depended on the kind of alloy used.

Besides the compositin of the metal, the corrosion products were analyzed. It was found that copper sulfates were always formed on the surface of the bronze, proving that this corrosion was due to the attack of the sulfur dioxide in the air. Tin is transformed into metastannic acid which is removed by the rain. The greatest damage resulting in pitting of the surface was found on bronzes rich in lead. The lead of bronzes is transformed into lead-sulfate, a white powder which is rapidly washed out by the rain. Among the substances on the surface of the bronze which are not due to the corrosion of the metal, quartz, iron hydrates and gypsum could be found. The gypsum is formed by the reaction of the sulfur dioxide with calcite, one of the main constituents of the dust. When deposited on the bronze, the gypsum bakes together the corrosion products, the dust, and the soot rich in aggressive compounds.

Figure 9
Putto in bronze (1613) at the monument in front of the Town Hall of Munich: The surface appears deeply corroded when the soot is removed.

A microscopic examination of sections through the metal and the surface layer showed that the green patina and the red cuprous oxides which are sometimes present *do not* form a protective layer on the metal but penetrate as an aggressive front into the bronze. As a result of this erosion which proceeds slowly but surely, more than one millimeter of the surface of the bronzes from the sixteenth century are already lost. Therefore, there are no more traces of the cold work on the bronzes by punching and chasing which essentially determine their quality as masterpieces of art.

The black crust over the metal and the green patina consist of alternating dark and light layers, the dark ones marking the soot from the periods when heating fuels are consumed, the light ones marking the dust from the summer.

Not only are the old pieces from the seventeenth and eighteenth centuries affected by this process of corrosion but also the monuments of the nineteenth century and the modern bronzes of the twentieth century. One sculpture by Arp, exhibited at Essen, had to be taken into the interior of the museum when deep holes appeared on the points where flakes of soot had settled.

Knowing the composition of the bronzes, the attacking compounds of the air and the process of corrosion, we have been able to develop methods for conservation and preservation. Coatings of oils, waxes and synthetic materials can be applied to recently cast or cleaned bronzes. Oils and waxes have the disadvantage of being effective for only a short time. The great personal and

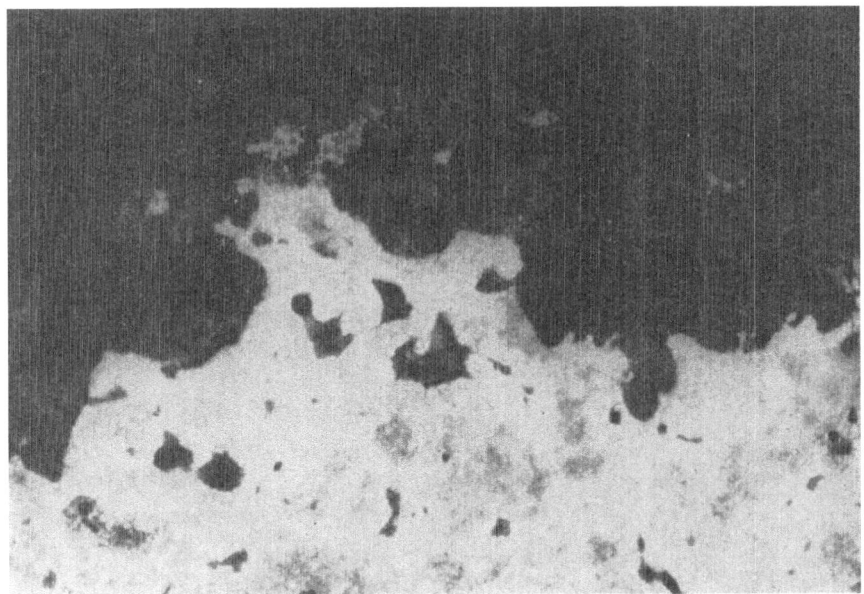

Figure 10
Microscopic examination of polished sections shows the deep corrosion of the metal (light), covered by soot and corrosion products (magnitude $50 \times$).

financial effort demanded by this type of preservation is the reason why many administrations are reluctant to simply rely on the resistance of the metal (which now has been shown to be a false assumption). The advantage of an application of oils and waxes is the natural and pleasant appearance of the surface attained by this treatment.

Coatings of synthetic materials have a considerably longer durability and a strong adherence to clean surfaces. For the protection of new bronzes the use of synthetic materials is better than that of the less resistant oils and waxes. On corroded bronzes, however, where the black and green products of the surface cannot be removed, synthetic materials may lead to an unpleasant change of their appearance.

Ninety plates, cast in bronze, were coated with different synthetic resins to find out which products among the great supply offered by the chemical industry for the protection of materials were best suited for our problem. The resins included 17 acryl resins, 9 polyurethan resins, 2 alcyd lacquers, 3 epoxy-resins, 1 silicon resin, 1 PVC/PVA lacquer, 2 combined lacquers, 5 lacquers of unknown base, 12 waxes and 4 oils.

A series of 10 bronze plates, covered with an artificial brown patina were also coated with different lacquers and waxes.

The ninety plates were exposed in the center of Munich where there is the highest concentration of sulfur dioxide (1965: 40 mg SO_2/100 m³, 1969: 15 mg SO_2/100 m³. The notable reduction of SO_2 in the air is due to the first efforts for

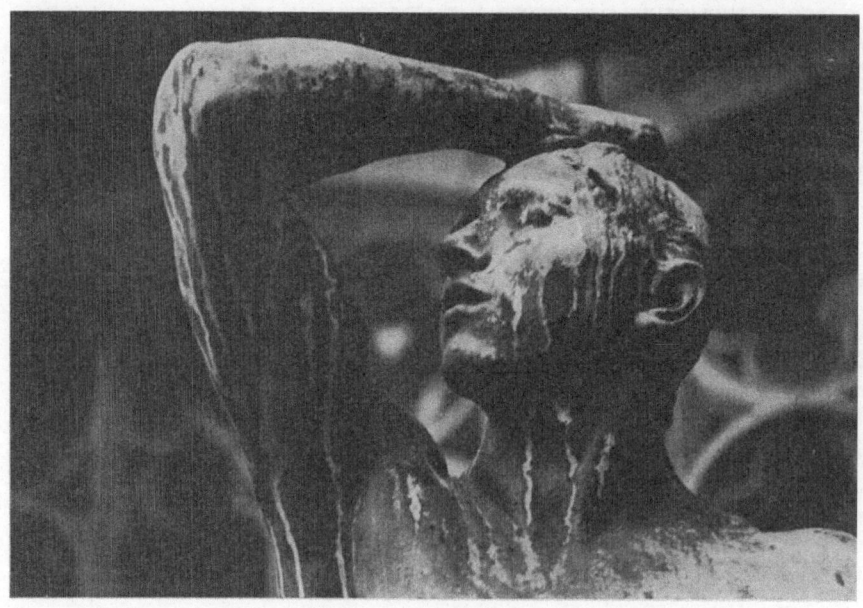

Figure 11
Sculpture in bronze by RODIN (1876) at the Wallraf-Richarz-Museum, Köln: Modern bronzes are also spoiled by traces of the rain through the layer of soot.

clean air in Munich by introducing electricity and natural gas instead of oil and coal for heating). The precipitation of dust amounts to 8 g/m² per month.

The results of the resistance of the test plates gives us hope that in the future damage to bronzes can be prevented. This together with the first restorations of seriously damages sculptures of the sixteenth century confirms that with our present technical knowledge, threatened works of art can be saved.

7   Damage in the Interior of Museums

The action of the polluted air is not restricted to objects exhibited outdoors, although that is where the attack of sulfur dioxide (which is only effective in the presence of water) is much more obvious. But dust and soot can enter those museums where the air is not cleaned thoroughly by filtering. The modern paintings in the Bavarian Art Gallery which were not covered with a protective varnish were found to have a strongly adhering layer of soot on the paint layer. The soot could not be removed without hurting the substance of the painting so it was decided from then on to put the paintings behind glass. Similar steps should be taken in many other museums in metropolitan areas around the world as soon as possible.

8   Documentation of Damage

Photographs were taken of all the objects where damage due to air pollution was found and were stored together with a written report about the material, corrosion products and methods of restoration. The photos are repeated regularly to see if the decay continuous on unprotected as well as protected objects and which method for conservation seems to be the best. All the data collected in these reports are stored in a preforated card system. By this documentation it has become obvious that many objects of great importance have suffered irreparable damage or are entirely lost. Only the close cooperation of scientists and restorers along with energetic efforts for clean air can help to maintain our works of art as essential withnesses of our culture.

New Economical Aspects of Air Pollution

by ERICH HOEDL
Institut für Makro- und Strukturplanung, Technische Hochschule Darmstadt,
Federal Republic of Germany

Abstract

Traditional economic theory is inadequate for the solution of air pollution problems. If environmental policy is guided by economic theory, pollution will be 'repaired' instead of prevented. Thus, on a societal level environmental policy will be irrational. The shift from economic theory *to* political economy *makes possible the integration of economic and technological aspects as well as the imperative cooperation between natural and social sciences. However, the political economy of air pollution can actually only be outlined to a limited degree. The tendency toward 'over-industrialization' and 'over-organization' as well as the concept of economic efficiency are still to be investigated in terms of political economy. The solution of pollution problems according to ecological principles has to take place within a global socio-economic strategy by* problem-oriented *planning processes on the level of active decision-making.*

1   Introduction

Economic theory has increasingly to cope with more than 'purely' economic problems as the economic dimension of society becomes more strongly interwoven with both sociological and technological aspects. Even economic problems (inflation, etc.) cannot be solved without taking into account the political and social processes. On the other hand, the growing *economization* of society gives rise to certain – most implicit – views, which contend that societal problems could be solved by a simple *reorientation* of economic activities.

The outstanding example of this perspective is the present strategy of solving environmental problems in western countries [1]. The governmental activites are limited to environmental legislation, in which standards of pollution and minor governmental controls of polluting activities are stipulated. Standards and controls apply mainly to the *consequences* and not to the *causes* of pollution. Additionally, the government establishes antipollution taxes and duties in order to impose adequate environmental response. The reorientation of the production processes under the new constraints is left entirely to industry.

The causes of pollution are to a large extent due to purely economic activities. Public demand for increasing production – which itself is strongly influenced by the advertising of the producer – is orientated toward narrow economic rationality, i.e. economic efficiency.

Since generally, economic efficiency is only concerned with market phenomena, it does not respond to ecological principles. As a result environmental disruption occurs [2]. *Economic theory* has never dealt with the impact of production and consumption on nature. The latter was only considered as a production factor. As air and water were not yet scarce, they were classified as 'free' goods (price=0). With the growing industrialization, prices were assigned to these resources, but only implicitly. Certain goods in areas with unpolluted air became more expensive than elsewhere (housing, etc.)

Considering air as a separate commodity is the essence of the actual economic discussion on air pollution [3]. According to traditional economic thinking, its price is the sum of the costs of 'production' of clean air. Inherent in this perspective is production by society as a whole. Society has to be considered as being composed of at least two groups, consumers and producers, and both contribute to this 'production'. The possible contribution of the consumer is, however, limited to a change of demand and its consequent behavior after the consumption of commodities. Contrary to this, the producer can take an active part in changing its technology and production program. As the producer also determines the possible choices of the consumer, he is playing the major role – although he is bound to consider existing technology and knowledge as well as the demand structure. Another difference exists in the remuneration: consumer and producer both get a clean environment but only the producer's income is directly affected by it.

Implicitly this perspective promotes the establishment of a new industrial sector which produces antipollution technologies (filter, etc.) in the short run and new technologies (electrocar, etc.) in the long run. Thus, only industry is providing a solution for the problem. The previously mentioned legal and organizational measures of government are instigating the strategy for solution.

Economic theory, due to its perspective that clean air has be 'produced', calls for an extension of the industrial sector: As there is a demand for a new *good*, additional *production* is considered to be necessary. The theory does not refer to existing production.

A more penetrating investigation of this concept for solution leads to questions, which seem to be important in an increasingly interdependent society:

a) What are the fundamental assumptions in the economic solution concepts and what are their implications for society?

b) What problems arise during the process of solution and can they be solved within the parameters of a rational strategy?

c) Can we design an alternative for the existing strategy? What conclusions can be drwan for economic theory and air pollution in general?

These questions are beyond the problem of partial solutions and are intended to outline an integrated perspective of air pollution.

It seems to be relevant to interpret air pollution not as a sum of partial failures of technological and social activities, but as a result of group interests and the distribution of socio-economic power. The framework presented in this study is intended to give an alternative to the discussion of pollution in terms of industrialization and population growth, which themselves can be explained historically and in terms of interests. The integration of differentd isciplines in working out partial solutions by means of simulation, optimizing methods, cost-benefit-analysis, etc. [4], or by an interdisciplinary planning approach are still in the beginning stages. They seem, however, to be more advanced than the problem of a socio-economic strategy on the level of our economic system as a whole.

2   Economic Theory and Technology

For the pollution problem no new concept has been proposed by economic theory so far. Until now only the existing tools have been applied to the new problem. Clean water is produced by water-management-system, garbage is used as a production factor (recycling) and finally, devices (filters, etc.) are installed between the source of pollution and the environment in order to produce clean air.

There are two fundamental concepts in economic theory:

a) The application of an optimizing method [5] (in most cases Linear Programming), in which the pollution standards appear as additional constraints, therefore reducing the level of production in order to protect nature.

b) The concept of input-output-analysis [6], extended by antipollution activity, the level of existing production remaining mainly unchanged.

Whereas in the second concept the extension of the activities is made explicit, the first concept might seem to propose a reduction of the national growth rate. As Linear Programming applies only to the individual firm in 'market economies', the national growth can be maintained by other activities or new firms. The reduced activities of the actual individual polluter are compensated by other firms, among which are the producers of antipollution devices. A growth of GNP is proposed within both concepts: explicitly within the rather descriptive model [concept (b)], implicitly within the rather normative model [concept (a)].

Within these concepts of growth, two strategies of pollution control can be applied:

Strategy I:

The production of clean environment starts *after* pollution, which is caused by production and consumption processes which have already disrupted the

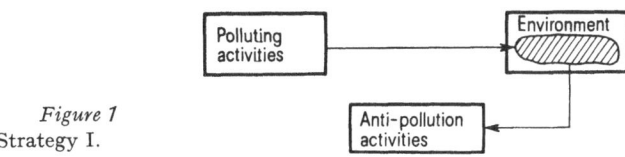

Figure 1
Strategy I.

environment. A typical case is garbage, which has to be collected after consumption. The same strategy can also be applied for air and water.

Strategy II:

Production and consumption cause pollution, but it is *reduced* before it enters the environment. A typical example is the installation of filters.

Figure 2
Strategy II.

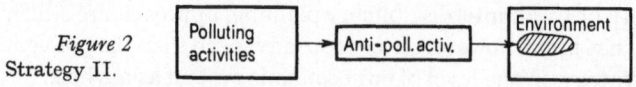

Both strategies are consistent with the two previously mentioned concepts of economic theory. In this respect economic theory proposes additional activities, allowing industry to choose between the two strategies. Economic theory does *not* refer to the strategy chosen: there is no link between economics and technology. The cooperation between economists and engineers has not yet been made explicit except in simple cost considerations.

3   The Process of Solution

The process of solving environmental problems is based on the principle of causality: Everyone who pollutes has to compensate for the damages.

This principle applies to *consumers* as well as to *producers*. The first can only prevent damages, whereas the latter can prevent or repair damages. If a polluted environment already exists, industry has to repair the damage and can choose in all other cases between preventing or repairing, according to its own preferences (profit, etc.).

For some firms both alternatives may turn out to be expensive and even lead occasionally to a shut-down. But industry as a whole may grow because the government must define pollution standards which still allow the supply of goods. At this stage an analysis of the power position of industry will show how far strategy I or II can be followed. In either case it would mean a growth of the new antipollution industry.

The application of the causality principle can be discussed from a technical, financial and organizational point of view. Strategy I gives rise to many problems in finding out the reference points between the damages and the causes. These reference points have to be measured in ecological and financial terms. Their organization has to be worked out on the local, national and international level. This organization seems to be easier in strategy II because there is a well-defined relationship between causes and consequences. The same applies to the measurement and evaluation of the measures taken. Assuming that all the *technical* aspects could be effected, there is still the question of *financing*. The different power positions of consumer and producer allow the latter to call for government subsidies or an increase of prices. Therefore the relation between industry and state results in increasing *organizational* activity of the state, which has to coordinate aspects of income distribution and adequate production by industry to meet consumer needs.

On the societal level the process of solution in both strategies discussed turns out to be expensive in three aspects:

Technical: The measurement of pollution requires organization and financing for instruments and labor. The evaluation demands empirical studies and criteria for the imputation of costs.

Financial: The distribution of costs is not based on theoretical studies, but is the result of bargaining processes. Environmental investments and distribution effects diminish economic welfare.

Organizational: Technical and financial aspects need organizational devices at all levels of government. Data collecting, information systems, and environmental planning are to a large extent social costs of private production and consumption.

The consequences of an environmental policy according to strategy I and II result in enormous costs for the *public administration*. These costs arise as long as industry can exclusively decide according to its individual preferences. In the present situation, government depends on industry in solving environmental problems. Only industry can solve the problem and government has to put forward conditions (standards, subsidies, etc.) which guarantee sufficient profits.

Private industry by its very nature can only produce under these conditions, implying that government has to solve all problems not solved by industry. The whole management of the environment is left to the state, which has to establish an adequate organization. In terms of social goals, the behavior of industry leads to 'over-industrialization' which government cannot prevent. In addition, governmental activities result in an 'over-organization'.

Within the transition period from a polluted to a clean environment the antipollution policy leads to:

a) 'over-industrialization': The 'production' of pollution is compensated by the 'production' of a clean environment. They are not harmonized on a societal level.

b) 'over-organization': The government has to take measures on all levels of administration to substitute for the former functioning of *competition* as a regulative device. The consequences of the autonomy of firms have to be compensated by the state.

Over-industrialization' and 'over-organization' are consequences of the isolated strategies of the industry and the state. As there exists no common theory and no real cooperation, each continues in its own direction and working against the other. A variation of 'countervailing power' [7] may arise; this means that solutions are reached only according to the power positions and in no respect by rational discussion.

If this process continues, there will be no change, even after the transition period, and production will still not be done according to ecological principles. The production sector may remain composed of pollution-producing and pollution-repairing industries. 'Over-industrialization' would then be institutionalized.

4   The Strategy of Preventing Pollution (Strategy III)

The demand for a commodity *clean environment* is the result of the production program, the consequences of consumption, and the production technologies which are not in accordance with ecological principles. Therefore, a change of technologies has been proposed. As economic theory is not related to the technological aspects directly, economists support this strategy only in terms of the argument regarding cost. The strategy of preventing pollution is adopted on a large scale by those firms, which have to pay for the damages they cause. This tendency is reinforced by industries developing new technologies, which in their turn can make profits from the pollution prevention in the long run.

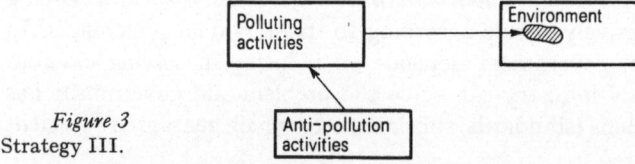

Figure 3
Strategy III.

An outline of the process of solution would show similar organizational problems as are found in strategies I and II, but strategy III would not result in 'over-industrialization'. The causes of pollution would be minimized and therefore 'repairing' the environment becomes unnececssary to the same degree.

Today, the policy is not orientated toward such a strategy of prevention, because there are *vested interests* in the existing production structure and possible high profits in the antipollution industry in the short run. Even some presently existing new technologies (electrocar) are not yet used in practice, the main reason being not the lack of technological production know-how, but the powerful position of the established industries which would eventually be affected by the new product.

If the strategy of prevention were applied beyond technological change, then an analysis of the relation between society and nature would have to be worked out. Those 'rules of the game' [8] which lead to disruption would have to be identified. This problem of a rational and democratic steering of society has at least an informational, motivational and organizational aspect [9]. The last aspect particularly needs increasing investigation, because out of this research more knowledge concerning the interests of the different social groups could be acquired. Environmental disruption is a consequence of the organizational structure of the society. In particular, the freedom of industry in choosing the production program, forces the government to establish an expensive control organization. But even by this measure the state might not be successful. The identification of the main power positions is a precondition for rational environmental policy.

5   Conclusion

The answering of the three questions seems to be a task for some kind of 'Macro-Sociology', which in an economized society would have to be a *political economy*. The current environmental policy is a reorientation of economic activities supported by the government. But also *economic theory* supports or is neutral vis-à-vis that solution. By this, economic theory is in favor of the present power distribution and scientifically without any relation to other social or natural sciences. Instead of the application of traditional economic theory the integration of both natural and social sciences could start with the problem of pollution on a societal level and within the *political economy*. From this long-term view, the specific problems of air pollution would be treated as a problem of interdisciplinary planning [10]. Such activities always pose technical, financial and organizational problems; they are by their very nature mixed strategies of repairing and preventing pollution. However, they would not result in *over-industrialization* and *over-organization*.

There should be no illusion about the success of this strategy. Even the modification of ownership and social planning in socialist countries has not been able – until now – to prevent pollution [11]. To a large extent it might be the total economization of society which brings about disruption. If the economic dimension of society is the dominant feature, it is also the dimension by which social control can be exercised effectively. The better the economic position of industry, the higher its social influence. In this sense, the request for control and power is the request for economic efficiency, which itself *must* lead to disruption. In an economized society the struggle for social power must lead to environmental disruption.

The investigation of economic efficiency can only be done by a political economy, which also has to specify the relation between economy and technology in order to allow cooperation with ecology. The switching from *economic theory* to *political economy* clears up strategical problems on the level of society and contributes to environmental planning on the level of active decision-making.

References

[1] G. KADE, *Ökonomische und gesellschaftspolitische Aspekte des Umweltschutzes*, Gewerk-schaftl. Mh., Köln *5* (1971).

[2] K. W. KAPP, *Environmental Disruption: General Issues and Methodological Problems*, Proceedings of International Conference Environmental Disruption (Ed. S. Tsuru, Tokyo 1970).

[3] R. M. SOLOW, *The Economists Approach to Pollution and Its Control*, Science, Wash. *173* (1971).

[4] E. HÖDL, *Techniken zur Umweltplanung*, Stadtbauwelt, Berl. *33* (March 1972).

[5] A. V. KNEESE and B. I. BOWER, *Managing Water Quality: Economics, Technology, Institutions* (Baltimore 1970).

[6]   W. W. LEONTIEF, *Environmental Repercussions and the Economic Structure*, Proceedings, op. cit.

[7]   J. K. GALBRAITH, *American Capitalism, The Concept of Countervailing Power* (New York 1952).

[8]   J. ROBINSON, *Economic Philosophy* (Middlesex, Engl. 1968).

[9]   E. NEUBERGER, *Libermanism, Computopia and the Visible Hand*, Am. Econ. Rev. *1966* (Paper and Proceedings).

[10] G. KADE, *The Economics of Pollution and the Interdisciplinary Approach to Environmental Planning*, Int. Soc. Sci. J., Paris *XXII* (1970).

[11] E. ALTVATER, *Gesellschaftliche Produktion und ökonomische Rationalität* (Frankfurt–Wien 1969).

Air Pollution, Air Conservation and International Economic Relations

by CHRISTOPH ANDREAS UHLIG,
University of Zürich, Switzerland

Abstract

Very little consideration has been given to the influences of air pollution and air conservation on international economic relations. Differences in the international treatment of air pollution will surely exist. Concerning the economic measures for air conservation, the application of rules and regulations seems to have an effect. A very simplified economic model shows that internationally differing stringency of measures for air conservation can induce various international economic flows. These economic movements have 'undesired' ecological results, which can be avoided only by additional flanking measures.

1 Introduction

For a long time now, economists have been painting a golden age born of economic growth. Today, however, extremely critical by-products of a fast and mostly unbroken development of the western world are increasingly overshadowing the more than two decades of economic growth-enthusiasm. Obviously, the threat of self-destruction of mankind caused by the environmental catastrophe is one of the most disastrous of these by-products.

Throughout the world during the last few years, the problems of environmental destruction have frequently been discussed. Yet economists did not seem to be exceptionally disturbed [1]. The discussion usually took place without the participation of economics as a science, due mainly to the assumption that problems of the human environment lie outside of the field of economic investigation. In addition, very little consideration had been given to the possible changes of the existing economic balance of power resulting from air pollution on the one hand, and of the widespread enforcement of pollution control measures on the other [2]. Nevertheless, it is important that changes in the economic balance be considered, as they may feed back to political decisions and social structures and may also influence the introduction of necessary measures against air pollution and hence the ecological situation.

The purpose of this short paper is to explore, from an economic point of view and in the most general terms, some aspects of the connection between air pollution and international economic relations. In a first section, we will look at some of the reasons for differences in the international treatment of air pollution. After that, we will endeavor to demonstrate the economic methods mostly proposed for air conservation. Subsequent to the discussion of some of the economic and ecological consequences of the mentioned international differences, some final remarks will be made. Although this paper treats *international* economic and ecological aspects, some results may be applicable to *interregional* and *intersectoral* problems.

2   Differences in National Policies

Various considerations can lead to the conviction that the countries fighting air pollution will do so in a different manner within the boundaries of their respective national sovereignties. We will state two forms of differentiation: First, there is some probability that the *stringency of measures* for the conservation of the atmosphere will differ internationally. Second, the *application of economic methods* used in conservation will differ.

Let us look at some reasons for the differentiation of stringency. Two reasons seem to be important:

a) the problem of subjective evaluation and the conflict of social goals,

b) the differences in the actual state of pollution.

The natural sciences are able to determine objectively the *changes* to the environment caused by human actions, but they are not able to determine environmental *damages* or *gains* just as objectively. Characterizing an environmental change as an environmental damage, however, is possible only on the basis of subjective evaluation. These subjective evaluations also influence the formulation of the goals in respect to the environment. The vague goal 'environmental protection', for example, implies the furtherance of positively evaluated changes in the environment as well as the hinderance of the negatively evaluated changes. The goals concerning the environment are not isolated from the other goals of society, but rather are linked together by various complementary and competing connections. Hence, the attainment of a goal (G_1) will promote the attainment of a complementary goal (G_2), and hamper that of the competing goal (G_3) (see Fig. 1). In the case of competing goals, there is an additional connection: the consequences of the attainment of one goal (G_1) on the competing goal (G_3) may feed back to the formulation of the former (G_1) (see Fig. 2). The competition of these two goals requires an evaluation of the importance of

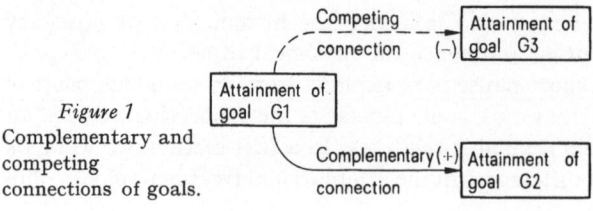

Figure 1
Complementary and
competing
connections of goals.

the goals in order to decide which will be preferred. This relative evaluation of goals within the hierarchy of goals raises the crucial question of which values and whose values should be relevant, and, depending on the political decision-making process, really are relevant. The result is a clash of interests between the different groups of society. Therefore, it may be difficult or even impossible to achieve a consensus of opinion within just one nation alone. Among sovereign states, differences in the mentioned relative evaluation of goals are bound to exist and consequently, international differences in respect to the stringency of measures for air conservation and the corresponding differences in the stress of the atmosphere will occur.

The stringency of norms for air conservation seems to differ for another reason. In the extensive ecological literature on environmental problems, a general line of thought can be followed: It is said that nature can assimilate a lot of the emissions caused by human actions. An increasing amount of emissions, however, will strain this assimilative capacity. Beyond a certain critical point, the assimilative processes will begin to deteriorate, and will finally break down. Hence, it is concluded that air conservation measures will differ inter-nationally, according to the actual degree of strain on the air's capacity: Countries with insignificant pollution and an undisturbed state of nature may find it convenient to adopt less stringent measures. Strong exploitation of the natural capacity, on the other hand, will necessitate extremely stringent measures.

The problems of subjective evaluation and the conflict of social goals will also influence the choice of economic methods to combat air polluting activities. As each method will have a different effect on each particular social group, the choice to be made will primarily be influenced by the groups and their specific interests.

It is impossible to discuss here the wide variety of economic methods and their effectiveness on combating air pollution. Hence we will have to be satisfied with a short review. We will distinguish three kinds of alternatives, which can also be combined among one another:

a) taxation of the polluters,

b) direct bargaining between the polluters and the polluted,

c) creation and application of rules and regulations to combat air pollution.

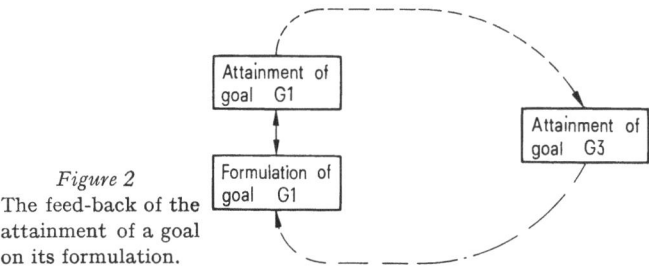

Figure 2
The feed-back of the attainment of a goal on its formulation.

Attainment of goal G1

Formulation of goal G1

Attainment of goal G3

The classical economic solution to problems like air pollution is the taxation of the polluters. The taxes are created by a central body as the price to be paid for polluting activities in order to reduce these activities. But the basic problem is still the question of where and in what way these indirect taxes should be imposed. Recent research has proposed the following solution: When pollution occurs on the *production* side, then a production tax should be imposed. In the case of pollution through the *consumption* of a commodity, however, a consumption tax is preferable [3]. If we assume constant consumer preferences and an independent supply (!), the size of the tax will depend on the price-elasticity of the demand function as well as on the degree of pollution-reduction desired. Estimating all the elasticities in order to determine the tax level raises numerous problems, to say the least. Other aspects of this method will have to be considered as well. Taxes will reduce the amount of polluting activities on which they are levied, but they will not eliminate them. In the case of a growing economy the taxes have to be constantly increased (according to the income-elasticity of demand), in order to keep polluting activities at the desired level. National and international monetary policy should be adapted to environmental policy, as the restrictive effect of taxes will be neutralized by creation of money. The problem of distribution also arises. The decision on who may pollute and who not and to what an extent depends on the purchasing power of the persons or groups concerned. Indirect taxes will have regressive effects if one assumes a constant consumer structure. Assuming, however, that the consumer structure will change, then the consumers with little purchasing power will withdraw from the circle of consumers. Only those with a high purchasing power will remain and consume the good concerned. The crux of the problem is that air pollution effects not only the polluter but other parts of society as well. The consumers who withdrew from the market suffer not only for no longer being able to consume a certain good, but also from the effects of the air pollution still remaining. Therefore, the levying of taxes to combat air pollution will most likely be impracticable due to the enormous economic problems which will arise and can even be unwanted because of the social effects it causes.

The second alternative is bargaining between the polluters and the polluted on the matter of pollution and its elimination. This may occur on a voluntary basis or under supervision of the courts, based on the assumption of a perfect system of property rights, where no polluting activities without compensation are permitted. Unfortunately, such a system of property rights is not possible when a good like air is concerned. Air is not attributable to any particular group or person, and hence no property rights can exist. Consequently, in the case of air pollution, this model is not applicable. But even when a good can be clearly attributable to a certain person or group, the unequal distribution of bargaining-power will make the realization of this solution highly questionable.

The third alternative is the establishment of a central body to formulate rules and regulations on the use of the air. Here, the pollution-problem is combated either by prohibiting or by restricting the sources of air pollution. We can assume that the rules and regulations will concern

- specific production processes,
- the consumption of certain products.

This solution is sometimes favored because it is supposed to give clear results.

3   Economic Consequences of a Regulatory Policy

We will now explain briefly the most important economic results of air conservation when enforced by the use of rules and regulations. This study is for the most part comparative-static. On a *short-term* basis we can assume constant preferences and constant technology.

Prohibiting rules and regulations in the production sector mean that the goods concerned will now no longer be produced in the same way as before. Likewise, the regulation of consumption means that the affected commodities will now no longer be consumed in the same way as before. The result is a decrease both of production and consumption. In order to maintain the existing level of economic activities, the polluting production activities in a *closed economy* will have to be substituted by nonpolluting production activities according to the rules and regulations laid down. Correspondingly, the polluting consumption activities will have to be substituted by nonpolluting consumption activities. These substitutions can be practicable *in the long run*. They may change the production as well as the consumption structure, since the permitted substitute activities will expand at the expense of the prohibited activities. The realization of these substitutes will inevitably raise the private costs of production and consumption, respectively.

The allocation of the productive factors in the case of air pollution is different to the case when air is not being polluted. The manufacturing branches causing the pollution directly through the production process, or indirectly through the consumption of the goods produced, may enlarge their outlets, since not all the costs of production and consumption have been evaluated. An effective application of the rules and regulations only allows activities which can bear all costs. It therefore eliminates – in the terms of welfare economics – the misallocation of productive factors due to air pollution.

The prevention of air pollution is supposed to have redistributive effects. Air pollution imposes heavy burdens on the economically weak strata of society especially [4]. It is this strata which has to bear, involuntarily and for the most part without protection, the inevitable passing-on of the effects of air pollution, since it is not in the position where its interests can be realized. The abolition of air pollution removes these burdens and therefore especially effects the weak parts of society.

4   A Simple Economic Model

In order to demonstrate some of the economic results of internationally differing stringencies of measures for air conservation, let us assume the existence of two trading countries, *A* and *B*. Trade is free and factor movements are possible. Country *A* decides to conserve the air by imposition of rules and regulations. Country *B*, however, takes no conservation measures. In this case of *open economies*, the situation changes because of the resulting international influences.

The application of the rules against pollutive production activities in country *A* eliminates certain domestic production processes, but the consumption of the commodities concerned still remains possible by the way of imports from country *B*. Consequently, in order to supply the market, the polluting production activities do not necessarily have to be substituted by nonpolluting production activities, for they can be substituted by imports of commodities which are no longer allowed to be produced in the same way as before in country

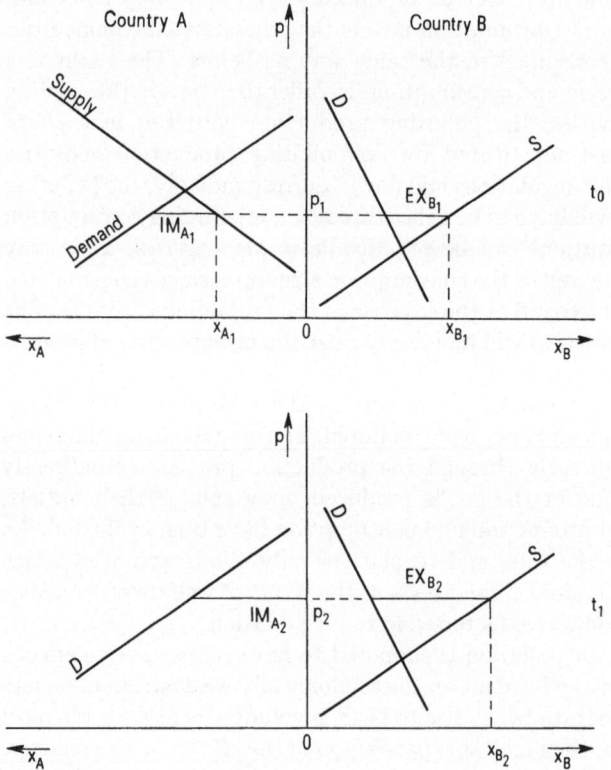

Figure 3

The substitution of polluting production activities in country A by increased imports –
country A initially importing (t_0 = *before*, t_1 = *after* the application of rules).

A. Should this be the case, and has country A already imported part of the goods in question, the quantity (and price) of imports will increase ($IM_{A_2} > IM_{A_1}$), corresponding to the respective price-elasticities. (For a very simplified demonstration see Figure 3: we measure the output x horizontally and the price p vertically. Line S shows that the higher the price, the higher the supply. Line D shows that the lower the price, the higher the demand.) If however, country A had exported part of its former production (EX_{A_1}), it will now import a certain quantity of the goods in question (IM_{A_2}), depending on the respective elasticities (see Fig. 4). In both cases, the result of such a substitution by way of importing, is an increase of the polluting production in country B ($x_{B_2} > x_{B_1}$). Consequently, country A decreases and country B increases its productive activities. The enforcement of (long run) substitute activities in country A will obviously tend to have comparable results.

The application of rules against polluting consumption activities in country A omits the domestic consumption of the goods concerned. However, (reduced) domestic production still remains possible for sale on the export market. The extent of reduction ($x_{A_2} < x_{A_1}$) depends on the respective price-elasticities. If country A had previously imported part of the consumed goods

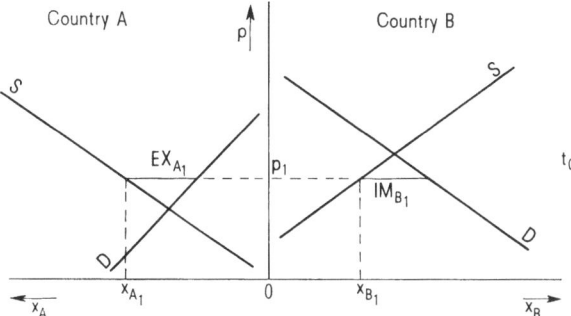

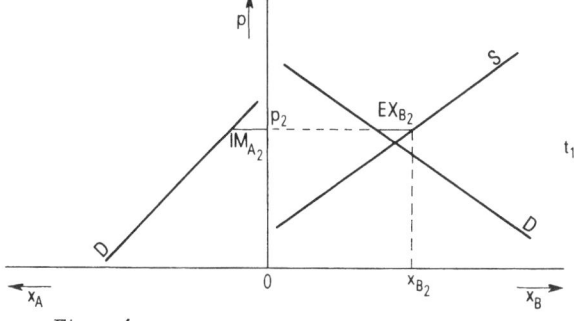

Figure 4
The substitution of polluting production activities by imports – country A initially exporting.

(IM_{A_1}), it will export the remaining production (EX_{A_2}) (see Fig. 5). However, if country A had exported part of its production, the amount of exports would increase ($EX_{A_2} > EX_{A_1}$), corresponding to the respective price-elasticities (see Fig. 6). Contrary to the production case, no perceptible evasion abroad is possible, as the omitted consumption can only be substituted in country A by domestic or foreign nonpolluting activities. Both country A and B will now suffer a reduction of the concerned production ($x_{A_2} < x_{A_1}$; $x_{B_2} < x_{B_1}$).

We must add, however, that the mentioned changes in exports and imports, as well as in economic activities will not be brought about in the degree indicated, as they will tend to be countered – for example by income effects caused by changes in the balance of payments (when assuming fixed exchange rates).

The use of rules and regulations will influence international factor movements. In the case of rules on production, the increased demand for exports from country B will improve the marginal-revenue-productivity of the factor capital. In country A, however, the marginal-revenue productivity of capital will tend to fall. Hence, the export of capital from (import of capital into)

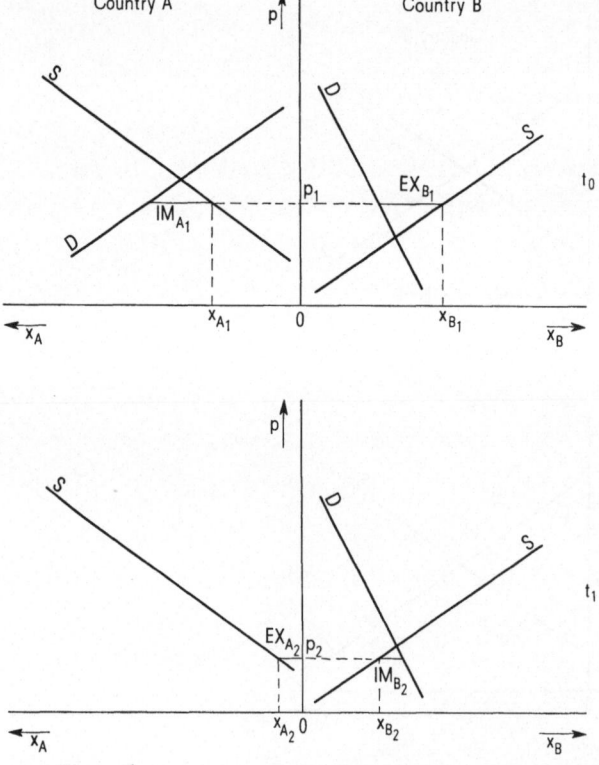

Figure 5
The abolition of the consumption of certain goods – country A initially importing.

country A will increase (decrease). In the long run, both countries will be influenced by these capital movements: country A towards an intensified decrease and country B towards an intensified increase in economic activities. – In the consumption case, the results are less clear-cut, since in both countries, the marginal-revenue-productivity of capital will tend to fall. The ratio of these reductions will determine the change in the flow of capital. If country A is influenced relatively more than country B, then the tendency will be an increase (reduction) of the existing export of capital from (import into) country A. In this case, the result will be an intensification of the decrease in economic activities in country A and a reduction of the decrease in country B.

The production factor labor will tend toward an increased immigration into and/or to a reduced emigration from the nonpolluting country A, due to the improvement of its quality of life. On the other hand, the change in economic activities will influence the remuneration of the factor labor, which may increase the emigration from and/or reduce the immigration into country A. The concrete socio-economic state of the production factor labor, the amount of changes in the quality of life, as well as its remuneration, will determine the prevailing tendency.

The use of rules and regulations will also alter the conditions on which international decisions on the location of firms are made. In the production

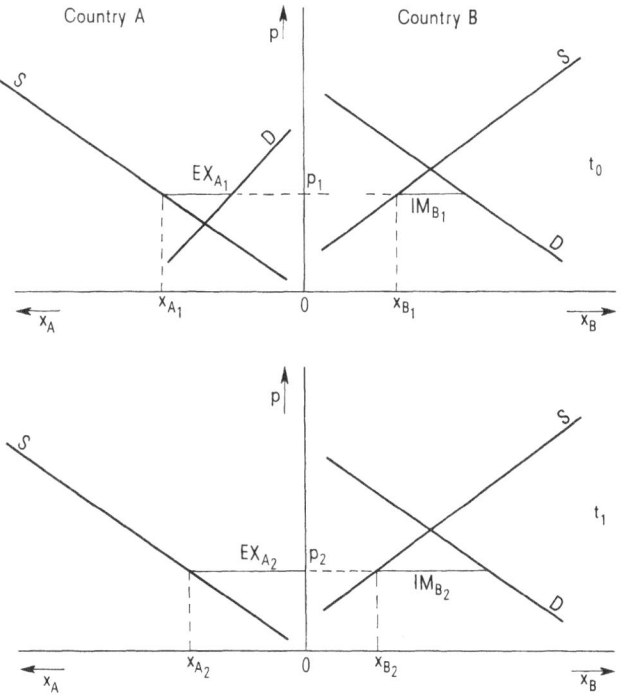

Figure 6
The abolition of the consumption of certain goods – country A initially exporting.

case, the maintainance of prohibited activities is possible only by transfer of production from country A to country B. In the consumption case, when we assume that production should be maintained, there will be a tendency to transfer production abroad, due to the change of outlets.

The economic short-term results of air conservation achieved through the application of rules and regulations in country A, with no such application in country B, can be summarized as follows: Should country A enforce the reduction of air pollution through rules and regulations on production, it will experience reduced economic activity. On the other hand, country B will experience increased economic activity. In the case of rules and regulations on consumption, country A, applying these rules, will experience a reduction of economic activity. But country B as well, applying no rules, will suffer a reduction of its economic activity. In both cases, changes in international capital flow, as well as in the international flow of products will set in. Decisions on the location of firms may be influenced decisively and movements of the labor factor may occur.

5   Ecological Results

The ecological results of the economic changes mentioned, due to differences in air conservation, can be stated as follows: The application of rules and regulations will influence
  – the state of pollution in country A,
  – the overall state of pollution,
  – the transfer of pollution across boundaries.

Due to the reduced strain on the atmosphere, the described air conservation measures in country A will lead to an improvement of its internal ecological state. Improvements on the overall state of pollution, however, are less clear-cut.

Should country A lay down production rules, and country B not follow suit, then the environmental improvement in country A will be countered by an added deterioration in country B: country A has exported its former pollution. The overall reduction of polluting activities will be smaller than the reduction in country A alone, since the polluting activities in country B will increase. If country A had the greater state of pollution, a tendency for equalization in both countries, rather than an overall reduction of polluting activities will result from the efforts in country A.

A different situation arises in the case of rules on consumption laid down by country A: the total quantity of pollution through consumption will decrease perceptibly. Only little evasion is possible, and the country practicing conservation cannot export its former pollution. Therefore, country A will not only achieve its aim of reducing the internal state of air pollution, but will also improve the overall state.

Air pollution in a certain country will not only influence that country's own state, but can also feedback to the other country by the transfer of pollution across boundaries, as the biosphere cannot be divided into national units. On the contrary: national actions concerning the environment are interrelated by the complex cycles of water, air, and minerals, all of which are not subject to the control of sovereign governments. These cycles may transfer pollution from the polluting country B to the nonpolluting country A. Consequently, the success of air pollution reduction in country A will depend on the state of air pollution in country B. This is especially the case, when rules and regulations are laid down in country A to prevent the spread of toxic and/or difficult-to-remove substances.

6 Concluding Remarks

It is not so easy to draw realistic conclusions from the model used. Only case studies, in which concrete assumptions on public and private action and reaction are made, can show realistic quantitative results, since numerous assumptions in the model – explicitly and implicity – do not, as a rule, conform to reality. We have seen that national measures against air-polluting production processes may influence not only the international flow of products and capital, but may also induce the transfer of polluting industries to countries with less stringent measures. In particular, multinational combines will be able to circumvent national regulations. These actions can be prevented, if the stringencies of national measures do not differ to such an extent that circumventions become profitable. Appropriate international agreements would be necessary in order to adjust the stringency of national methods, as well as the type of economic methods used. However, political reality demonstrates daily the improbability of such extensive international agreements. Another way to prevent circumvention is to prohibit the import of products manufactured by production processes not conforming to the importing country's rules. Prohibiting imports may lead to severe international problems, as this can be interpretated to be an interference in the sovereignity of other countries. Nevertheless it remains the only practical way to carry through an overall reduction of polluting production processes rather than a transfer of the ecological problem from one country to another.

We have also shown some of the influences that national measures against air-polluting consumption have on international economic relations. In this context, we must state that if is very problematic to continue exporting products which have been declared undesirable for domestic consumption. Logic would prohibit the export of these products, meaning that their domestic production would be stopped completely. A well-known example of the futility of isolated internal measures is the ban of the use of DDT in the United States, as decreed by the American Environmental Protection Agency. The reason given was the unacceptable risk to human life and the environment [5].

Notwithstanding, the ban does not apply to the manufacture of DDT for export.

To summarize, national measures against pollution cannot be restricted to domestic production and consumption activities alone: They must be complimented by flanking measures.

The environmental problem cannot be dealt with on the basis of the opinion that stringent air conservation measures are unnecessary in low-populated or economically less developed countries. The mentioned international economic movements may set in and last as long as the state of nature in these countries has deteriorated to such an extent, that the application of stringent measures would be indispensable and great efforts would have to be made to remove the results of pollution. Consequently, the effects of possible economic movements should not be considered as an assistance to 'the industrialization efforts of the receiving country' [6]. This kind of assistance will not solve, but will rather raise problems.

Let us make three final remarks: 1. The problems of pollution cannot be properly solved by partial and isolated consideration of environmental media like air and water. Pollution stresses a total ecological web and the organisms living in it. 2. The political decision-making process reflects not only the balance of power between social groups within one country, but also the type and the extent of the relationship between these groups and similar groups in other countries. 3. The existing economic tools to treat environmental problems are largely inadequate, as they ignore some very important facts: the interrelationships in complex ecological systems, the adjustments in dynamic economies, and the concrete socio-economic situations. Only a more integrated approach can lay part of the groundwork for abolishing the threat of self-destruction of mankind.

Acknowledgment

I am indebted to BEAT A. REGLI and HANS D. CHI, both at the University of Zürich, for their comments.

References

[1]   E.g., the most widely read economic textbook contains only a small chapter on pollution: P. SAMUELSON, *Economics*, 8th ed., New York 1970.
[2]   For exceptions see e.g.: GATT, *Industrial Pollution Control and International Trade*, Geneva 1971; S. F. MAGEE and W. F. FORD, *Environmental Pollution, the Terms of Trade and Balance of Payments of the United States*, in: Kyklos *XXV*, 101–118 (1972).
[3]   S. F. MAGEE and W. F. FORD, op. cit., p. 102.
[4]   K. W. KAPP, *Umweltgefährdungs als ökonomisches und wirtschaftspolitisches Problem* – Lecture held at the 1971 annual meeting of the Wirtschaftswissenschaftliche Gesc schaft für Oberösterreich in Bad Ischl.
[5]   Quoted by the Umwelt-Report *17/72*, p. 8.
[6]   GATT, op. cit., p. 23.

Trans-Frontier Air Pollution and International Law

by Peter H. Sand
Rome, Italy

Abstract

After critical analysis of the leading international reference case, the 1941 US-Canadian Trail Smelter Arbitration, an alternative legal approach to trans-frontier air pollution problems is outlined and illustrated by recent European experience. Instead of recourse to special international procedures, most of these problems can more efficiently be resolved by adapting local procedures so as to accommodate foreign parties. The basic remaining problem, which indeed requires new international rules, is identified as long-distance air pollution, especially pollution affecting environmental resources outside the area of national jurisdiction.

1 Introduction: The 'Trail Smelter Arbitration'

Air pollution is not merely a national problem common to most modern states – it also tends to become an international problem whenever its effects transcend national boundaries. Are there any international legal rules applicable to the problem?

The leading, and indeed the only widely-known, case in point is the 'Trail Smelter Arbitration' which settled a trans-frontier dispute between the United States and Canada in 1941. Canada was held responsible for damage caused to farms and forests in the State of Washington by SO_2 fumes emitted from a smelter located on Canadian territory at Trail, British Columbia, and consequently had to pay damages and take specified preventive measures. In a much-quoted sentence the tribunal, chaired by a Belgian lawyer, concluded:

'Under the principles of international law, as well as of the law of the United States, no State has the right to use or permit the use of its territory in such a manner as to cause injury by fumes in or to the territory of another or the property or persons therein when the case is of serious consequence and the injury is established by clear and convincing evidence' [1].

There is little doubt that the Trail Smelter case made good sense for the specific dispute at hand. It continues to be cited with approval in the two countries involved and abroad [2]. Its method of dispute-settlement, including the setting of effective emission standards and monitoring procedures, was a pioneering example of interdisciplinary cooperation between legal and scientific experts.

Beyond this, however, there has been a good deal of international legal myth about the Trail Smelter. To begin with, its diplomatic-political background and the degree of prior consensus were so unique that they will hardly ever be duplicated in a normal trans-frontier pollution situation elsewhere. Far from having been brought by action in an international court, the case was voluntarily submitted to arbitration by a special US-Canadian convention, which was concluded six years before the final decision [3] and which in turn stemmed from an earlier recommendation by the International Joint Commission in which both countries had been cooperating since the 1909 Boundary Waters Treaty [4]. This specific arbitration agreement, and not the general rules of international law per se, constituted the terms of reference of the arbitral tribunal so established. In fact, the tribunal was specifically directed also to consider such non-international sources of law as the US Supreme Court's decisions in interstate air pollution cases [5]. In view of these circumstances, the precedent value of the Trail Smelter Arbitration for other international pollution disputes is very limited, and even its above-quoted statement of what reads like a universal principle of international law has been dismissed as a mere *obiter dictum* in legal literature [6].

Yet we are not concerned here with niceties of legal reasoning, nor with the somewhat specious argument whether the rules of the case may be extrapolated as 'hard' precedent or as 'soft' analogy only [7]. The real question is whether this kind of dispute-settlement is at all relevant today; i.e., relevant and useful for the resolution of typical contemporary and future conflicts over international air pollution.

2   Alternative Approaches

The Trail Smelter case is by no means the only reported international settlement of a trans-frontier air pollution problem. Besides subsequent US-Canadian problems, in which the International Joint Commission played an important preventive role [8], there has been an exchange of diplomatic notes concerning the emission of offensive odours across the US-Mexican border in 1961, later settled by the 1963 Chamizal Convention [9]. A major dispute arose in 1957 at the French-German border concerning air pollution by a power plant on the French side, which led to considerable political agitation including a tax-strike in the German township affected and to formal questions raised in the federal parliament [10]; the matter was eventually resolved by financial arrangements for antipollution measures under the Saar Treaty [11]. In 1960, there were parliamentary questions in the French National Assembly concerning air pollution by a Swiss oil refinery across the border at Aigle–Collombey [12]; proposals to extend the powers of the Swiss-French Lake Geneva Commission to include air pollution matters did not materialize, though [13].

In 1967, a special French-German Commission was established to examine air pollution problems raised by the construction of a chemical plant in the

French border-town of Ottmarsheim. In a meeting at Paris in 1968, the Commission approved antipollution measures and agreed on monitoring arrangements [14]. New problems have also been reported from the border between France and Switzerland, following Swiss protests in 1969 against fumes emitted from a French plant at Annemasse [15].

Partly as a result of the growing number of these incidents, the Council of Europe in 1971 passed a resolution recommending 'that governments of member States of the Council of Europe ensure for the inhabitants of regions beyond their frontiers the same protection against air pollution in frontier areas as is provided for their own inhabitants. To this end they should in particular ensure that the competent authorities should inform each other in good time about any project for installations liable to pollute the atmosphere beyond the frontier. The competent authorities beyond the frontier should be able to make their comments on such projects. These comments should be given the same consideration and treatment as if they had been made by the inhabitants of the country where the plant is situated or proposed' [16].

This new 'localizing' approach may indeed offer a viable alternative to the 'internationalizing' approach of the Trail Smelter Arbitration. Obviously, had it not been for the accidents of political geography, the farmers aggrieved by the smelter at Trail or by the chemical plant at Ottmarsheim could have complained to their local authorities, sued in their local courts, and nobody in Washington D.C., Ottawa, Paris or Bonn would ever have heard of the case. As it happened, the mere existence of the boundary added a new factor and raised local problems to an intergovernmental level where a special international procedure had to be devised to solve them. Ever since Trail Smelter, therefore, international lawyers seem to have concentrated exclusively on the kind of clinical procedure required for this purpose. Yet if the only cause of the complication here (as compared to 'normal' pollution cases) was the boundary handicap, a more rational remedy would be to eliminate this handicap which distorted the equities and prevented the parties from treating the problem like a normal domestic one. Instead of internationalizing a local issue (via an enormous detour to the respective national capitals), a more economic solution would be to adapt local decision-making processes so that they can handle trans-frontier problems like ordinary local ones of comparable size.

To be sure, 'deflating' a trans-frontier air pollution problem requires a number of adjustments in the legal-administrative systems on both sides of the border. Considering the problems encountered in the United States even for those few air quality control regions which cut across state boundaries [17], it may be somewhat premature to call for 'international air quality control regions' already [18]. While there are precedents for direct (short-cut) consultations and voluntary cooperation for pollution control between local authorities across national boundaries [19], recourse to regular administrative machinery and to the courts raises special problems of standing to sue and choice of law for foreign parties. Thus, in a 1969 decision concerning the new Salzburg jet airport, the Austrian Supreme Administrative Court dismissed environmental objections raised by an adjoining German township across the border (Freilassing) as legally inadmissible on the theory that the Austrian Aviation Act ceased to be applicable right at the border and was not meant to protect foreigners [20]. This would confirm a view expressed earlier by a Belgian legal scholar as regards the limited scope of national air pollution laws [21], against a French jurist's opinion who favours the invocability of such laws for the benefit of victims on the other side of the boundary [22].

Perhaps the most interesting recent development in this respect is a 1957 judgment by the Court of Appeals of Saarbrücken (West Germany), particularly since the facts of the case are quite close to the Trail Smelter situation. It arose from the above-mentioned Saar boundary dispute over fumes and coal-dust emitted by an electrical power plant at Grosbliederstroff, set up in 1954 as a subsidiary of the state-owned French Lorraine Basin Mining Company. One of the victims in the small town of Kleinblittersdorf on the German side of the Saar river, owner of the 'Rebenhof' country resort, brought action against the French company in the local German court for damages to his agricultural and horticultural crops and to his recreation and tourism business. The Appeal Court held that under the rules of private international law as interpreted in Germany, the plaintiff had a choice to sue either under German law or under French law, whichever seemed more favourable to him. The German plaintiff opted for French law, because of its easier rules on proof, and was awarded delictual damages under Article 1384 of the Code Napoleon [23].

The point, then, is rather simple: Instead of blowing up trans-frontier air pollution cases to international size, other ways can be found to deflate and de-internationalize them so they can be handled in their proper local context. While this approach requires a number of adjustments between state laws, the social-administrative costs involved would seem to be lower than those of over-sized Trail Smelter procedures.

3   Conclusion

This does not mean that international legal rules will become superfluous for air pollution problems. On the contrary, they need to be developed to cover ground hitherto totally neglected. It will be noted that the Trail Smelter Arbitration only spoke about damage 'in or to the territory' of another State [24]. This *territorial* limitation and fixation is typical of classical international law; in fact, most textbooks of international law would list the Trail Smelter case under 'territory' or 'neighbourhood relations' [25]. The tendency parallels the property-fixation of classical private law, which refused for a long time to recognize any pollution damages other than those incurred by 'neighbours' (i.e., adjoining landowners [26]) – and it has been related psychologically to the 'territorial imperative' which may be part of the more beastly heritage of our species [27].

The result in the international pollution sphere is twofold. First, none of the existing bilateral arrangements for trans-frontier air pollution control can cope with problems of long-distance pollution such as the acid rains from Western and Central Europe observed in Scandinavia [28]. Secondly, traditional international law can give no remedy at all for environmental damage which is inflicted not on the territory of an individual victim, but 'merely' on common resources such as the oceans, the Arctic, or the global atmosphere. This situation perfectly illustrates what has been described as 'the tragedy of the commons' [29].

One important step forward has been the 1972 Stockholm Declaration on the Human Environment. For the first time, the international responsibility of states is extended from the old territorial types of damage to environmental damage in 'areas beyond the limits of national jurisdiction' [30]. Admittedly, legal implementation of this extension (for which most credit is due to the Canadian delegation and its concern with Arctic pollution) will not be easy in practice, and still requires solid interdisciplinary research. Here, and not in trans-frontier neighbourhood relations, is a genuine need for new *international* law on air pollution control.

References

[1] Second decision, as reported in: Am. J. int. Law *35*, 716 (1941); text of the first decision ibid. *33*, 182 (1939); see also the technical expert opinion by DEAN and SWAIN, *Report Submitted to the Trail Smelter Arbitral Tribunal*, US Bureau of Mines, Bulletin No. 453 (1944).

[2] E.g., see READ, *Trail Smelter Dispute*, Can. Yb. int. Law *1*, 213 (1963); and WHITEMAN, Dig. int. Law *6*, 253 (Washington 1968).

[3] Convention for the Settlement of Difficulties Arising From Operation of the Smelter at Trail, British Columbia; 13 April/3 August 1935; as reported in: Am. J. int. Law *30*, Supplement, 163 (1936).

[4] Docket No. 25 of the IJC; see WHITEMAN, Dig. int. Law *3*, 840 (Washington 1964).

[5] E.g., the early (1907) decision in *State of Georgia vs. Tennessee Copper Company and Ducktown Sulphur, Copper and Iron Company Ltd.*, 206 U.S. 230, 237 U.S. 474, which held at p. 477: 'It is a fair and reasonable demand on the part of a sovereign that the air over its territory should not be polluted on a great scale by sulphurous acid gas, that the forests on its mountains, be they better or worse, and whatever domestic destruction they may have suffered, should not be further destroyed or threatened by the act of persons beyond its control, that the crops and orchards on its hills should not be endangered from the same source.'

[6] See BILDER, *Controlling Great Lakes Pollution: A Study in United States-Canadian Environmental Cooperation*, Michigan Law Rev. 70, 492 (1972).

[7] See also RUBIN, *Pollution by Analogy: The Trail Smelter Arbitration*, Oregon Law Rev. *50*, 259 (1971).

[8] See REMPE, *International Air Pollution: United States and Canada: A Joint Approach*, Arizona Law Rev. *10*, 138 (1968); JORDAN, *Recent Developments in Environmental Pollution Control*, McGill Law J. *15*, 279 (1969); and STEIN, *The Potential of Regional Organizations in Managing the Human Environment*, Woodrow Wilson Center Environment Series No. 202, p. 26, 58 (Washington 1972); for the text of the 1966 air pollution reference, see Am. J. int. Law *61*, 112 (1967).

[9] As reported in: WHITEMAN, Dig. int. Law *6*, 256 (Washington 1968).

[10] Bundestag, question No. 347 submitted by 30 deputees on 28 March 1957.

[11] Following the recommendations of a special international commission (created on 12 December 1957), a jointly-financed filter system and two 140-meter chimneys were eventually installed in 1959–60, at a total cost of 15 million francs, and allegedly reducing the air pollution by 90%. However, on 2 April 1971 a new complaint by residents of the same township was formally submitted, stating a new substantial increase in air pollution from the French plant, due apparently to gradual deterioration of the filters and to the company's reluctance to make further antipollution investments in the plant which is scheduled to be closed down within a few years.

[12] Text of the reply by the French Ministry of Foreign Affairs in: J. off., No. 14781, p. 1389 (1962).

[13] Ann. fr. Droit int. *8*, 1010 (1962).

[14] Procès-verbal de la réunion franco-allemande concernant l'usine P.E.C.-Rhin à Ottmarsheim, Paris (Ministère des Affaires Etrangères), 19 January 1968.

[15] Revue gén. Droit int. publ. *73*, 185 (1969); see also LUCCHINI, *La pollution du milieu naturel*, J. Droit int. *96*, 1085 (1969).

[16] Council of Europe, Committee of Ministers, Resolution (71) 5, adopted on 26 March 1971: *Air Pollution in Frontier Areas*, nationally implemented, e.g., in German Bundesgesetzblatt (1971), part II, No. 35, p. 975 (24 July 1971); see also the Council's *Declaration of Principles on Air Pollution Control*, Resolution (68) 4, adopted on 8 March 1968, title II, chap. 7; for earlier activities see ADINOLFI, *First Steps Toward European Cooperation in Reducing Air Pollution: Activities of the Council of Europe*, Law Contemp. Probl. *33*, 421 (1968).

[17] See LIEBER, *Controlling Metropolitan Pollution Through Regional Airsheds: Administrative Requirements and Political Problems*, J. Air Pollut. Control Ass. *18*, 86 (1968); ZIMMERMANN, *Political Boundaries and Air Pollution Control*, J. Urban Law *46*, 173 (1968); and GREEN, *State Control of Interstate Air Pollution*, Law Contemp. Probl. *33*, 315 (1968), for a critical assessment of the role of interstate compacts.

[18] See LEE, *International Legal Aspects of Pollution of the Atmosphere*, Univ. Toronto Law J. *21*, 203 (1971).

[19] E.g., some of the joint initiatives in the Detroit/Windsor area in the framework of the US-Canadian IJC; or implementation of the Lake Constance Convention between Austria, Germany and Switzerland.

[20] Verwaltungsgerichtshof Vienna, cases No. 233/69 and 314/69, judgment of 30 May 1969 summarized in: Öst. Juristenztg. *25*, 305 (1970); see also the critical note by SCHREUER, *Zur verwaltungs- und völkerrechtlichen Problematik des Salzburger Flughafenfalles*, ibid. *26*, 542 (1971).

[21] P. DE VISSCHER, *La protection de l'atmosphère en droit international*, Rapports Généraux au VIIe Congrès International de Droit Comparé (Brussels 1968), p. 339, No. 4.

[22] KISS, *La protection de l'atmosphère en droit international*, Etudes Droit Contemp., p. 374 (Paris 1966); see also SAND, *Internationaler Umweltschutz und neue Rechtsfragen der Atmosphärennutzung*, Z. Luftr. Weltraumrechtsfr. *20*, 111 (1971).

[23] *Poro vs. Houillières du Bassin de Lorraine (HBL)*, judgment of Oberlandesgericht Saarbrücken on 22 October 1957 (case No. 2U45/57) as reported by LEVY, in: Neue Jurist. Wschr. *11*, 752 (1958); following this test case, a number of related claims were settled by HBL with other German plaintiffs.

[24] Supra note [1].

[25] On the concept of 'international neighbourhood law' in this context see ANDRASSY, *Les relations internationales de voisinage*, Recueil Cours Acad. Droit int. *79*, 77 (1951/II); THALMANN, *Grundprinzipien des modernen zwischenstaatlichen Nachbarrechts* (Zürich 1951); SÖRENSEN, *Principes de droit international public*, Recueil Cours Acad. Droit int. *101*, 194 (1960/III); VON DER HEYDTE, *Das Prinzip der guten Nachbarschaft*, Verdross-Festschrift, p. 133 (Vienna 1960).

[26] On these 'mind-forged manacles of law' see especially SAX, *Defending the Environment: A Strategy for Citizen Action*, p. 125 (New York 1971).

[27] Compare ARDREY, *The Territorial Imperative: A Personal Inquiry Into the Animal Origins of Property and Nations* (New York 1966); and EHRENZWEIG, *Psychoanalytical Jurisprudence* (Leyden 1971).

[28] See the special report prepared by the Swedish government for the Stockholm Conference, and the study carried out by the Norwegian Institute for Air Research on behalf of the OECD Air Management Research Group; see also MÜNNICH, *Report on the Pollution of the Sea Through the Atmosphere*, Joint Group of Experts on the Scientific Aspects of Marine Pollution (GESAMP), Doc. No. III/12.5.4 (1971); and LUNDHOLM, *Interaction Between Oceans and Terrestrial Ecosystems*, in: *Global Effects of Environmental Pollution* (Singer, ed., Dordrecht 1970), p. 195.

[29] HARDIN, *The Tragedy of the Commons*, Science *162*, 1243 (13 December 1968); see also CROWE, *The Tragedy of the Commons Revisited*, Science *166*, 1103 (29 November 1969).

[30] United Nations Declaration on the Human Environment, adopted on 16 June 1972, Principles No. 21 and 22; UN Monthly Chronicle *9* (No. 7), 86 (July 1972).

Air Pollution Control in the Developing Country

by MICHAEL G. MCGARRY
Division of Environmental Engineering, Asian Institute of Technology,
Bangkok, Thailand

Abstract

The technological aspects of air pollution control are only one component of an abatement program. This paper pertains to the developing country, its air pollution problem; motivation towards pollution control; the paucity of expertise, research and educational programs in air pollution, and administrative and enforcement difficulties encountered by the developing country implementing air pollution control programs.

1   Introduction

Scientists are often criticized for their esoteric activities and apparent inability to evolve useful solutions to real problems. This paper relates to implementation of air pollution control measures and suggests why it is that, although technical solutions may be available, they are not utilized in the field. It covers the many steps from recognition of the problem to enforcement of air pollution regulations. It is also about developing countries – specifically those of South East Asia, their situation with respect to air pollution and their need to control it. Areas of interest within this context are the motivation to control air pollution in the developing country, existing research and monitoring facilities, personnel, education, administration, law enactment and air quality standards enforcement. All these aspects differ in character from those in the developed or industrialized state; these differences have an important and often negative influence over the effectiveness of air pollution programs in the developing country. This paper is not intended to be a treatise on the implementation of effective air pollution programs but rather a discussion of differences between the developed and developing countries with respect to air pollution control.

2   The Developing Country

It is characteristic of development that one city dominates the national economy. Apart from the city states of Singapore and Hong Kong, which can hardly be termed as developing, the developing countries of Asia are agriculturally based. It is in the city that air pollution problems first become significant and it is here that population growth is most rapid. Relative to rural population growth, urban increases are not exceptional. However, in absolute numbers of urban population the current level of growth rates are about 5% for cities of greater than 20,000 population – or a doubling of population every fifteen years. This rate has rarely been exceeded by the developed country city.

'The projected increase in the urban population of the developing countries in the 40 years from 1960 to 2000 is over one billion, more than four times the increase in the previous 40 years and about three times the total urban population of the developed world in 1960.' World Bank Group [1].

Some of the largest cities are more than doubling their population in the space of a decade. This is characteristic of the primate city which is often several times the size of the next largest. Centres such as Manila, Bangkok and Djakarta are economically more attractive and expand at a greater rate than provincial centres. Unfortunately, the major cities are often ill-planned and are not capable of handling the influx. Transport systems are based on relatively few main thorough-fares which are not well interconnected by feeder roads. As a consequence the transport systems are characterized by choked main arteries supported by a maze of ill-used, poorly connected, narrow alley-ways. Unrelieved urban population pressures place serious strains on urban transport systems.

Reasons for migration to the cities include population pressure on the land, better earning capacities and services in the cities and improvements in life-style. If development is to proceed, this trend to urbanize will continue. There is a near linear relationship between the ratio of urban to total population and the GNP per capita (from India with 19% urban population and a US$ 110 GNP/cap, to Spain with 59% urban population and a US $ 800 GNP/cap).

Economic dominance of the primate city in the developing country is exemplified by automobile registration data. In 1969, the 3.8% of the national population making up Djakarta owned 36.6% of the country's automobiles. Comparative 1966 figures for Bangkok were 7.9% and 72.8%, when the city's motor-vehicle population was 157,200; four years later it had jumped to 303,700. Vehicles are relatively old, athough body maintainance is good, engine up-keep and tuning are consistently poor. The densest traffic is invariably found in the commercial centres. In contrast to Western cities, multiple land use prevails; commercial areas are also the most heavily populated residential districts which exacerbates the public health problems caused by air pollutants emitted from poorly maintained engines on traffic jammed thorough-fares.

The air pollution problem is largely caused by the internal combustion engine. Carbon monoxide has been recognized as the single most important air pollutant of Bangkok, McGARRY [2]. Source emission CO levels averaged at 4.8% for petrol driven cars, a similar situation persists in Seoul, South Korea,

CHUNG [3]. Ambient levels in Bangkok's dense commercial section have been observed as being often above 13 ppm (eight-hour daylight average.) This is well above the US eight-hour Federal Standard set at 9 ppm.

3   Motivation for Air Pollution Control

In contrast to the developed countries, civic action pressure groups and public demand for environmental improvement are not strong. The widely publicized environmental deterioration of the industrialized nations has, however, had its impact on developing countries. Some reactions to this publicity have been negative on the basis that developing countries should industrialize first and worry about the environmental consequences later. Much of the positive action has been motivated by personal objectives: government officials finding it politically astute to join the 'environmental bandwagon'. Whatever the motivation it has not found its origin in the common man. Most environmental campaigns of the developed world have used democratic political systems to achieve their ends. These means of stimulating government action are not available to the people of developing South East Asia. The people most affected by such adverse consequences of industrialization as air pollution are not used to voicing public opinion nor do they have the political structure through which they can make claim to their governments.

In the developing country, a greater proportion of wealth is shared by a relatively small number of people, this is in contrast to the more even distribution of wealth in the developed country. It is unfortunate that those receiving direct economic benefit through development are not those suffering from the environmental damage that it causes. Likewise, it is unfortunate that those persons controlling the economic sector are extremely influential in government. Naturally, there is a reluctance to impose any environmental constraint on commercial ventures which would reduce their profit margin. The costs of using clean air are not being recognized by those making use of it as they are not normally those affected by its pollution.

4   Research and the Government Scientist

Not all technology is appropriate for import from the developed countries. It is true that analytical equipment may be purchased abroad but solutions to air pollution problems and air quality criteria have to reflect indigenous conditions and therefore must be developed in the country for which they are intended. Air pollution is however a new subject for most of these countries and government personnel who are capable of understanding even the bare fundamentals are very scarce. Consequently, there are very few effective research programs in the area. The practice of providing scholarships for overseas training is common. Unfortunately, there is a tendency in such programs to concentrate on specific aspects of air pollution related to the industrialized state, whereas the developing countries need personnel with a broad understanding of problems and solutions in air pollution pertaining to their situation. Governments are forced to look to existing personnel, who have inadequate experience and education.

The government scientist is normally limited to purely specific and technical matters. The structure of authority in developing countries is hierarchical; the government scientist is reluctant to enter into the political arena, to make statements against government policy, to embarass the government or to oppose politically influential pressure groups. It is therefore common to find the scientist isolated from reality, unwilling to make a forceful contribution and fearful of action which might be taken against his career should he step out of line.

Promotion is based on seniority, there is therefore, little incentive for research or innovation. No recognition is given to achievement as may be measured in terms of publication. Motivation to do anything other than that which is ordered by higher authority is lacking. Salaries are discouraging, there is often a factor of more than three between government and private sector salaries. This puts further strain on the government's very limited access to qualified personnel. There appears to be little government direction given on broad science policy. What little money is available for research is most often divided between too many research projects without adequate scrutiny or priority rating. As a natural outcome, there is a continuing reliance on overseas research and the direct import of often inappropriate technology from abroad.

5   Education

In general, there is a lack of awareness of public health consequences of air pollution. This subject is rarely given attention within formal education programs. Mass education by radio and television tends to be political and cultural in nature. What coverage exists is usually given in local journals and is most often tainted by emotional overtones and seldom technically correct. The urbanite has come to regard his situation in a fatalistic manner, as if deterioration in the quality of his environment were inevitable – a part of his way of life. The effects of air pollution are not immediate or drastic enough to make him react against it.

The patterns of higher education in developing countries commonly follow the classical structure which emphasizes division of academic disciplines. One of the primary causes of the slow development of environmental science and engineering has been the inadequacy of the educational system to bridge the gaps between disciplines. There are very few cross linkages between universities or even government departments. These will be necessary in future; solutions to environmental problems demand interdisciplinary approaches and personnel capable of drawing on several fields of study. The training of government personnel in environmental science and engineering through extension services is not available – there is little motivation to improve educational status once full-time education is completed.

6   Administration

With the recognition of air pollution problems and the desire to solve them comes the need to establish new governmental agencies for implementation of control programs or to allocate the responsibility to existing government departments. Several Western governments have responded to public pressure by forming new agencies or ministries of the environment with wide ranging authority over environmental protection activities. In the developing countries, public pressure is, however, seldom strong enough to result in the formation of a new ministry. Responsibilities for environmental control are most often divided between existing departments, authority being vested in those departments which are closely associated with the pollutant sources. Unfortunately, existing departments have territorial fears and ambitions which interfere with implementation of the program. Authority may be vested in the traffic police department for control of vehicular emissions; other departments which should be involved, such as those of health, town planning and transportation are often excluded or are not sufficiently in communication to contribute to the control program. The hierarchical structure of authority within a department also has its drawbacks. Vertical comunication is poor; the department head or manager has mixed feelings of fear, and ambition: fear of working in a new area in which his subordinates may have superior technical ability, and ambition to gain authority over new programs. Existing departments have too much straight-line momentum to solve new problems; personnel are geared to solving old ones, with the result that old functions are continued under new headings, MEYBOOM [4]. The demand for qualified personnel with expertise in new fields with strong inter-disciplinary overtones is not met. The most likely outcome is the formation of a committee; the result being the division of responsibility for decision making between a large number of concerned officials.

Thailand's response to publicity and pressure both at home and abroad was to form an Environmental Quality Control Committee comprising the representatives of twelve organizations. This was superseded after the Stockholm conference by a National Environment Council of superior authority, chaired by the retired chief of police. No action arm has yet been established. By government decreee, a limit was set on particulates emitted by automobiles; this was a political decision having little effect on public health. Carbon monoxide (CO) levels are currently not controlled although a limit of 4% may be set in the near future. The effect which control of emissions to below 4% will have on street CO levels is largely a matter of conjecture as no correlation between source and ambient levels has been established.

7   Air Quality Standards and Enforcement of Regulations

Autocratic government has the advantage of being able to circumvent the otherwise lengthy procedure of legislation. On the other hand, abbreviated procedures may lead to enactment without adequate scrutiny. With insufficient technical input, there is a tendency to over-simplify the content of regulations and to import environmental quality standards from the industrialized states which do not reflect the physical and social conditions of the developing country. There is an unmet need to establish the relationship between pollutant source levels and local ambient quality. Persistent traffic choked conditions of the primate city are quite different from those of the higher capacity streets and thorough-fares in the developed countries. To meet the same physiological requirements of ambient air quality, source limits must necessarily be more stringent in the developing countries. Engine cycles used in testing vehicles must also reflect local traffic speeds and conditions. Finally, predictive relationships between source and ambient levels of pollutants are required to ensure that the emission levels chosen are neither too severe nor too lenient.

It is common to use the automobile licencing agency and traffic police to enforce air pollution degrees or legislation. However, in the developing country there is a paucity of expertise in both; educational levels are low, training is often poor and an understanding of what is being enforced and why is commonly non-existent at the enforcement level. The equipment used for analysis of source emissions is commonly in a state of ill-repair. As often as not, there is no local dealer available to standardize or repair analytical equipment. As equipment is scarce, the breakdown of an essential component can paralyse the pollution control program for months while the instrument is freighted overseas for repair. Poor maintainance of equipment gives rise to unreliable measurements of source levels. Often, at the heart of the problem of enforcement lies the willingness of police to accept bribes; salaries are low and the temptation to use authority to raise personal income is high. The public is often unaware of pollution control parameters and the means by which improvements may be made. As a consequence of this and other reasons given above, the public becomes confused, skeptical and even fearful of pollution abatement programs.

8   Summary

Pollution control is a costly business, to make things worse the benefits accrued are most often neither visible nor measurable in monetary terms. It is natural, then, that profit making ventures should gain priority in allocation of economic resources over environment protection programs.

The polluters are most often those components of an economy upon which national development relies. Industry, whose expansion is encouraged in order to provide better standards of living, is a recognized source of pollution. Urban centres, prerequisite for development, are characterized by domestic pollution, low environmental quality and social upheaval. Government is faced with legislating against and constraining the very sectors it relies upon for economic development. It must ensure that a compromise is reached whereat society benefits from an improved environment and industry and commerce are not unduly restricted to the point which is detrimental to society as a whole.

In many countries, environmental control programs are proving to be successful, many have led to the formation of such powerful bodies as the Ministry of the Environment, England, Environment Canada and the Environmental Protection Agency of the US. In such cases, the governments concerned have been able to respond realistically to the emotional pleas and sometimes exaggerated warnings of the conservationists. In this respect, South East Asia has a long way to go, although the formation of government infrastructure to cope with the complex problems of environment control is already under way.

Air pollution problems are centred in the cities and caused by vehicular emissions in the denser residential-commercial districts. Asia's cities are expanding rapidly. Congested traffic and narrow streets give rise to carbon monoxide levels which are well above accepted maximum levels set in the developed countries. Motivation to control pollution has not been derived of the common urbanite who remains in relative ignorance of the adverse effects of his environment. In contrast to the developed countries, the third world's political systems do not encourage civic action pressure groups. Any initiative towards environmental control is taken at a relative high level in government and is usually in response to adverse publicity.

Research, monitoring and enforcement programs suffer as a consequence of insufficient expertise in the field of air pollution. Career advancement in the civil service is still largely based on the seniority principle, there is little incentive to innovate. The creation of pollution control programs is beset with territorical fears and ambitions of existing government departments, the outcome is often the formation of another committee with limited authority and no action arm. There is also an undesirable tendency to rely upon the industrialized states for research in the field and to borrow inappropriate air quality standards from abroad to be set up as standards in the developing countries.

At the heart of the problem lies inadequate education of government personnel. There is a tendency for pollution control programs to be flash-in-the-pan affairs; the government gaining merit through the resulting favourable publicity but allowing the effort to subside soon thereafter. As a consequence, the public often becomes confused, skeptical and even fearful of abatement programs.

As described above, air pollution abatement programs are fraught with seemingly insurmountable difficulties, the future appears to be bleak. Yet, in many ways there is a similarity between the conditions existing today in the third world and those in the developed states only a few decades ago. The paucity of education and expertise in the field will be relatively straight forward to overcome. It will be some time however, before the governments of these countries are sufficiently convinced of the benefits of environmental protection programs to provide funds and administrative infrastructure for their effective implementation.

References

[1]   World Bank Group, *Urbanization: Sector Working Paper*, (Washington D.C., June 1972).
[2]   M. G. McGARRY, *Surveys of Environmental Pollution in Thailand*, Thai Management Association Seminar, Bangkok (1972).
[3]   YONG CHUNG, in: BAE SHON, SUNG BAIK CHANG, MYONG CHO YOON and SOOK PYO KWON, *Study on Auto Exhaust Gases of Various Vehicles*, J. Korean mod. Med. 7, No. 5 (1967).
[4]   P. MEYBOOM, *Science in a Changing Environment*, Science Policy Branch, Policy, Planning and Research Service. Environment Canada, Ottawa, January (1972).

Eco-Engineering

This part of the book is dedicated to new approaches and ideas on how to improve today's environment: the development of a cure for our sick planet. The goal of research and development in the field of environmental pollution is the control, and, ultimately, the elimination of the sources of pollution. A necessary step along the way, however, is the development of suitable techniques for monitoring the presence of pollution already in the environment. The objective is not simply to register the presence of such pollution; our eyes and noses are in most cases already sensitive enough indicators of that. Rather, we seek to carry out quantitative measurements over a period of time, so that the effects of various abatement strategies can be evaluated. It is important to have accurate three-dimensional maps of the local density of various atmospheric pollutants, and their evolution in time, in order to test dispersion models for atmospheric transport. The proposals of tax penalties on emission of such substances as sulphur dioxide provide the legal and economic incentive for the development and deployment of sensitive, accurate monitoring schemes that will pass the test of evidence in a court of law.

Various monitoring systems are presently being used all over the world, and new systems are being developed. One of the most promising research projects in this field is being undertaken by Prof. J. STEINFELD and B. GREEN at the Massachusetts Institute of Technology in developing a laser ranging system for monitoring pollution levels. The spectroscopic properties of molecules which constitute air pollutants provide and ideal method for monitoring the trace concentrations with which they often occur in the atmosphere. Conventional spectroscopic techniques are generally unsuitable, hovever, for sampling the long paths over which such concentrations must be measured. It is for this reason that there has been a great deal of interest in the applications of coherent optical sources to such measurements.

If different system are being used or if measurements in different countries are to be compared, an internationally accepted calibration system must be developed. Miss J. M. DESHPANDE of the Central Public Health Research Institute in Nagpur, India, has been working on an X-ray-fluorescence system and believes it may well be adopted as an international calibration method, as it is simple to handle and is subject to minimum errors. She also gives her impression of the air pollution problem in developing countries, a slightly contrasting view to McGARRY's contributions and thus inviting further discussion. Although we must realize that we can not solve the environmental crisis just by developing new technical antipollution equipment, we must not forget that the new technology will necessarily play a vital part in our complex efforts. I did not wish to edit a book describing only the newest devices for cleaning polluted air. I believe, however, that a detailed look in one field of present air pollution technology is necessary in order to give a correct over-all picture of air pollution research. Combustion of fossil fuels is today the major source of air pollution; the dominant pollutants are hydrocarbons, carbon monoxide, soot, sulphur dioxide and nitrogen oxides. The first three result from incomplete combustion and sulphur dioxide is a result of the presence of sulphur compounds in the fuel, and its reduction is the problem of the refinery. Nitric oxide emission, finally, is due to the oxidation of atmospheric nitrogen in the combustor. The formation of photochemical smogs is caused by this emission. With a more complete understanding of the reactions of this nitric oxidation, the actual combustion process could be designed so as to reduce this emission substantially. At the University of Leeds in England a research project under D. THOMPSON is investigating this polluting process of nitric oxide emission. In his contribution he deducts from his theoretical and experimental results how this pollution can be reduced by technological alterations of present combustion processes, and what future power systems seem to be most acceptable judged by this environmental constraint. It is an instructive example of a true 'eco-engineering' approach.

Monitoring Atmospheric Pollution Levels by Laser Ranging Systems

by B. D. Green and J. I. Steinfeld
Department of Chemistry, Massachusetts Institute of Technology, Cambridge, Massachusetts, USA

Abstract

Atmospheric pollution levels can be monitored by several optical ranging schemes using lasers. These schemes include Raman backscattering, resonance fluorescence backscattering, and long-path resonant absorption. The three schemes are compared and evaluated. The long-path absorption method appears to have the greatest sensitivity. A system for monitoring atmospheric NO_2 with tunable organic dye lasers at levels of less than one part per million is suggested.

1 Introduction

Today most monitoring of atmospheric pollutants is done by the classical chemical methods of air-sampling followed by 'wet' bench-top analysis. In the United States, for instance, the National Air Pollution Control Administration (a division of the Environmental Protection Agency) currently monitors 42 atmospheric pollutants in this way, including 21 elemental species, nine gaseous molecules, particulates, and a variety of other components [1]. Such methods, while reliable, are tedious, insensitive to rapid concentration fluctuations in time and space, and poorly suited to rapid, automatic gathering of the large quantities of data required as input to computer models of atmospheric pollution.

In principle, optical methods of analysis should be capable of providing the desired information. Conventional spectrophotometric methods, however, are far too insensitive to be applied to the detection of trace contaminants widely dispersed in the atmosphere. The development of spectroscopic techniques using high-power, tunable lasers promises to change this situation. Such lasers can be tuned so that their output coincides with a resonance in a particular species whose atmospheric concentration is desired, free of interference from other species present; the power available makes it possible to range through long atmospheric path lengths.

Four basic air pollution monitoring schemes, using laser sources, have been proposed [2, 3]. These are

1. elastic (Rayleigh, Mie) backscattering,
2. inelastic (Raman) backscattering,

3. fluorescence backscattering,

4. long-path resonant absorption.

The first technique, also known as 'optical radar' or 'Lidar' is most effective for the detection of suspended particulates, which are typically of the order of 1 μm in diameter. These particles produce a strong optical echo for radiation having a wavelength of similar dimensions, such as is produced by the ruby (0.6943 μ) or Nd:YAG (1.06 μ) lasers. This technique is not sensitive for specific atomic and molecular pollutants, and so will not be considered further. The other three proposed techniques will be discussed in the following sections, in terms of their sensitivity limitations, suitability for specific monitoring problems, and currently available optical technology.

2   Remote Raman Spectroscopy

Raman scattering results in a shift in frequency of the scattered light, induced by a change in vibrational and/or rotational state in the molecule from which the light is scattered. The process is shown schematically in Figure

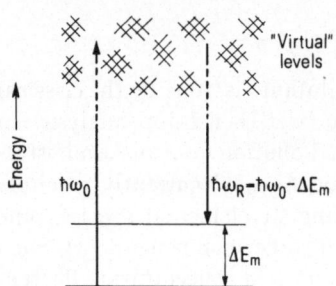

Figure 1

Principles of Raman spectroscopy. a) Light at frequency ω_0 is scattered at frequency $\omega_R = \omega_0 - \Delta E_m/\hbar$, corresponding to a transition ΔE_m in the molecule. b) Raman-scattered light from a polluted atmosphere. From [4].

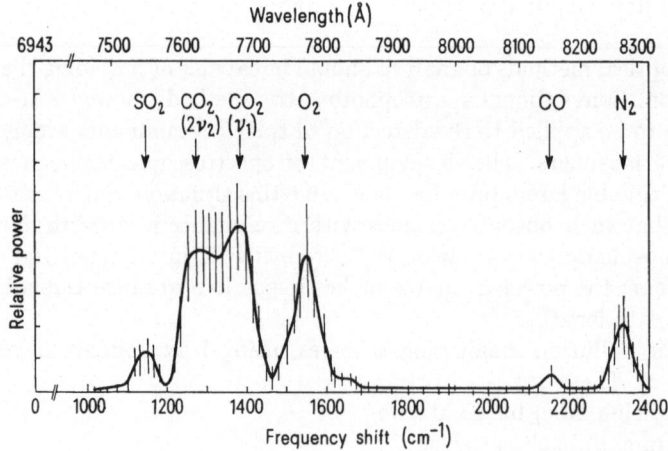

1a; an example of Raman-backscattered light actually obtained from a polluted atmosphere is shown in Figure 1b [4].

The Raman detection scheme possesses a number of obvious advantages, but one major limitation.

1. The wavelength of the scattered light is specific to each pollutant. Thus, a narrow-band filter can be placed in front of the detection systems which will select the Raman light for the specific substance desired, and reject both the intense elastically scattered light at ω_0 and much of the background sky radiation. Both are serious potential sources of interference in any backscattering detection scheme.

2. A single laser frequency can be used for monitoring all pollutant species. This permits the intense fundamental output of a ruby or neodymium laser to be used, without the need for additional optical components. As we shall see, however, some tuning of the output wavelenth may be required in order to overcome signal intensity limitations.

3. Depth resolution is possible by time-gating the returning Raman signal. Since light travels 3×10^{10} cm/sec (30 cm/nanosecond is a convenient metric), resolution is limited only by the effective bandwidth of the receiver system. But once more, intensity requirements will place a limitation on this bandwidth.

4. As with all backscatter techniques, the apparatus required is single-ended. Only a telescope receiver at the laser location is needed, making for a self-contained, portable monitoring device.

The intensity of the returning Raman signal, I, in photons/cm² sec, is given [2] by

$$I = \frac{N_0(R) T_1 T_2 c W_0}{2 R^2} \left(\frac{d\sigma}{d\Omega} \right)_{\text{Raman}} \tag{1}$$

where $c =$ the speed of light (3×10^{10} cm/sec) and T_1, $T_2 =$ atmospheric transmission factors at ω_0 and ω_R, respectively, which are of order unity. If we assume a concentration of pollutant N_0 (R) at 1 ppm in the atmosphere ($\simeq 3 \times 10^{13}$ molecules/cm³), a range of detection $R = 1$ km $= 10^5$ cm, and the number of photons in a 1-joule laser pulse, $W_0 \simeq 10^{18}$ photons, and since $(d\sigma/d\Omega)_{\text{Raman}}$ is not greater than 10^{-28} cm²/molecule steradian[t] even under favorable conditions, we find an intensity $I = 4 \times 10^3$ photons/cm²sec. Using a 10×10 cm aperture telescope in the receiver, we may expect to collect 4×10^5 photons/sec; but if we want as much as 30% resolution of the 1-km range, i.e., 300 m, we can collect photons only during the time it takes light to travel 300 m, i.e., $\Delta t = 3 \times 10^4/3 \times 10^{10} \simeq 10^{-6}$ sec. In this interval, we can expect to collect one photon, on the average, for each laser pulse sent out.

This presents a severe limitation on the general applicability of this method. In photon counting, we expect the noise due to statistical fluctuations to be of the order of the square root of the counting rate. In order to achieve even 10% relative accuracy from one measurement to the next, we require a signal/noise ratio greater than 10, which in turn implies a counting rate greater than $100 \, \text{sec}^{-1}$. Averaging of 100 successive signals would provide the necessary improvement in signal/noise; but the repetition rate of 1-joule Q-switched lasers is low (~ 1 pulse/minute), so that the time resolution would be lost in the process.

It is therefore not surprising that only a few results in Raman sensing of air pollution have been discussed [5, 6]. In one experiment [6], Raman back-scattering from SO_2 was observed from the top of a smokestack at short range when pure sulfur was burned in the furnace. In this instance, at least, the experiment may have caused more pollution than it cured!

Resonance Raman. One way to increase the sensitivity of the Raman detection scheme is to operate the pumping laser near an absorption frequency of the molecule being sampled. By a resonance effect, the effective cross section ($d\sigma/d\Omega$) can be increased by a factor of 10^3 to 10^4 in this way. However, this approach requires that the laser be tuned to the proximity of a molecular absorption, such as the 3,500 Å band of SO_2 or the 4,000–6,000 Å band of NO_2. This means that tunable nonlinear optical elements have to be incorporated into the system. In addition to increasing the cost and complexity, this reduces the available pulse energy W_0 by as much as 90%. On balance, a resonance-Raman remote-ranging detection system appears marginally feasible, but not a generally applicable technique.

3   Resonance Fluorescence Backscattering

The way to increase the sensitivity limits of the backscattering technique is to measure not Raman scattering, but the emission of resonance fluorescence by the species being monitored. Fluorescence involves the absorption of a photon at a particular resonant frequency, followed by reemission of light after a mean delay τ_r, called the *radiative lifetime* of the system. The cross section for resonant absorption is typically of the order of $10^{-18} \, \text{cm}^2$/molecule steradian – approximately 10 orders of magnitude greater than for Raman scattering. Fluorescence can be observed both in the visible part of the spectrum, involving excitation of electronic energy levels, and in the infrared, involving excitation of vibrational energy levels.

There are, however, several limiting factors that must be considered when dealing with fluorescence. The most important of these is the *quenching* of the excited molecule by collisions during the lifetime of the excited state. Molecules that are quenched give up their excitation energy to the collision partner, with the result that the fluorescence is not emitted. The rate of quenching collisions is given by

$$Q = (\sqrt{2}) N \, (\sigma^2/Z) \bar{v}, \tag{2}$$

where N is the density of the quenching gas in molecules/cm³, σ^2 the kinetic collision cross section, and $\bar{v}$ the average molecular velocity. Z is the average number of gas collisions required to quench the excited species. In air at standard temperature and pressure ($N \simeq 3 \times 10^{19}$ cm^{-3}), the rate of gas-kinetic collisions ($Z = 1$) is approximately 5×10^9 sec^{-1}. The *quantum yield* of fluorescence is given by

$$\phi = (1 + Q\tau_r)^{-1}; \tag{3}$$

so if the quenching rate is much greater than the fluorescence decay rate, very little backscattered light will be seen.

For electronic transitions in the visible part of the spectrum, τ_r is typically in the range 10^{-6} to 10^{-8} seconds. Quenching efficiencies have been measured for a number of species involved in atmospheric pollution, including NO, NO_2 and SO_2 [7]. These quantities can show quite large variations, but most of the systems of interest have Z's of the order of 1–10. For the particular case of optically excited NO_2, $\tau_r \simeq 4 \times 10^{-5}$ sec and Z for quenching by N_2 is approximately 8, so that $Q \simeq 6 \times 10^8$ sec^{-1} and $\phi \simeq 4 \times 10^{-5}$.

For vibrational fluorescence in the infrared, τ_r is generally 0.01 to 1.0 second. Quenching by vibrational deactivation has been measured for a number of small polyatomic species [8], for which Z varies between 10^2 and 10^4. Water vapor is notoriously efficient in quenching vibrational excitation; since this is always present to some degree in the atmosphere, a reasonable minimum estimate for Q for infrared fluorescence would be 5×10^6 sec^{-1}. Thus, a quantum yield no greater than 2×10^{-6} should be expected.

In order to estimate the intensity of backscattered fluorescence, the resonant absorption cross section reduced by the quantum yield must be inserted in (1). Let us take $(d\sigma/d\Omega) \simeq 10^{-18}$ cm^2/molecule, $\phi = 4 \times 10^{-5}$, $N_0(R) \simeq 3 \times 10^{13}$ molecules/cm^3, and $R = 10^5$ cm. Since the laser must be tuned to a molecular absorption line, no more than 10^{17} photons/pulse can be expected if the optical conversion efficiency is 10%, and we obtain an intensity $I = 2 \times 10^8$ photons/cm^2 sec. Again, if we use a 100-cm^2 aperature to collect the fluorescence, we may expect to gather 10^4 photons in a 1-μsec interval (corresponding to a depth of resolution of 300 m). In this case, a 1% relative accuracy in the measured concentration could be obtained.

The infrared fluorescence case is not nearly so favorable. With the lower (2×10^{-6}) quantum yield for vibrational fluorescence, a returning flux of 10^7 photons/cm^2 sec may be anticipated. Infrared detection cannot count individual photons, but respond only to a net energy input. A flux of 10^7 5-μ photons/cm^2 sec corresponds to 3.6×10^{-6} ergs/cm^2 sec; thus, a 100-cm^2 detector could collect 3.6×10^{-11} watts. A detectivity greater than 3×10^{10} cm(Hz)$^{1/2}$/watt at 5 microns, with a bandwith greater than 1 MHz, would be required to register this amount of energy; only the very best liquid-helium-cooled semiconductor detectors approach this sensitivity [9]. It may be possible to improve this limit by the used of optical heterodyne reception [10].

One situation in which fluorescence backscattering is a much more promising possibility is the detection of metal atoms dispersed in the atmosphere. Atomic spectral transitions are very intense, so that the absorption cross sections are high ($\sim 10^{-14}$ cm²). By the same token, the lifetimes are usually of the order of 10^{-8} to 10^{-9} seconds, so that quenching is relatively less important; typically, $\phi \sim 0.1$ to 0.01. Under these circumstances, a return of 10^8 photons could be expected for an atomic pollutant dispersed at a level of one part per *billion* in the atmosphere, which is a typical concentration. Some specific applications are suggested in a later section.

4   Long-Path Resonant Absorption

A somewhat different approach to the measurement of dispersed pollutants involves monitoring not the backscattered radiation, but the transmission of radiation tuned to a resonant absorption frequency of the molecule. In this case, the relationship between light intensity and concentration is simply

$$\frac{I}{I_0} = e^{-k(\lambda)pl} \tag{4}$$

where I and I_0 are the received and transmitted laser pulse intensities at wavelength λ, $k(\lambda)$ the absorption coefficient at that wavelength, p the concentration of pollutant, and l the total optical path length. Absorption coefficients are typically of the order 0.1–1.0 (ppm-km)$^{-1}$, so that easily measurable optical attenuations can be expected. There are other processes besides resonant absorption responsible for attenuation in the atmosphere, of course; the method relies on making measurements at two neighboring wavelengths over which $k(\lambda)$ shows a large variation but scattering by particulates, fog, etc. is essentially the same.

The absorption technique has the advantage of making use of the full laser intensity, so that excellent signal/noise figures can be obtained. The

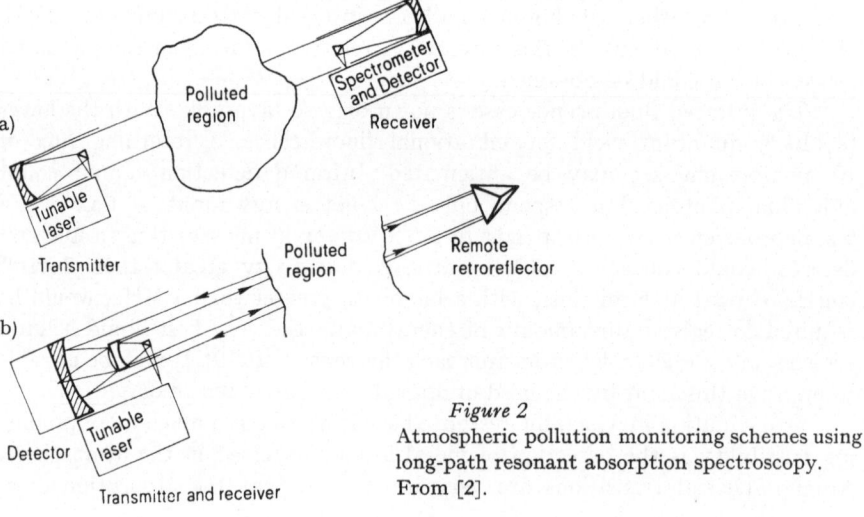

Figure 2

Atmospheric pollution monitoring schemes using long-path resonant absorption spectroscopy. From [2].

disadvantage is that depth resolution is lost, since the beam samples the atmosphere along the full length of its range. (We will show below, however, that it may be possible to incorporate depth resolution into an absorption scheme.) Two possible schemes for atmospheric monitoring are shown in Figure 2; the approach shown in Figure 2b is preferable, because both the source and the detector are installed at a single central location, while a number of passive retroreflectors can be placed around the periphery of the polluted region to give wide yet inexpensive coverage. Another approach is that of 'point sampling': namely, collecting gas samples and bringing them to an absorption apparatus set up in a laboratory. In such an approach, of course, the advantage of real-time spatial resolution has been lost.

There have been only a few reports of pollution monitoring using resonant infrared absorption spectroscopy [11]. Tunable diode lasers have been used to detect ethylene [11], SO_2 [12] and NO [13]. Since these lasers have a power output of only about 10 microwatts, they are pretty much restricted to point-sampling applications [14]. More powerful Raman spin-flip lasers, pumped by a CO laser, have been used to detect NO [15]; in this case, detection was by means of an acoustic signal generated in the absorbing gas, which can be measured only in a laboratory set-up after point-sampling. The CO laser itself can be made resonant with a transition in NO only if the transition itself is Zeeman-shifted by applying a magnetic field of several kilogauss to the NO molecules [16]; this, of course, cannot be done under atmospheric conditions. In a subsequent section, we examine the feasibility of monitoring atmospheric NO_2 by using a tuned organic dye laser operating in the visible part of the spectrum.

5 Tunable Radiation Sources

Both resonant fluorescence and absorption measurements require a laser that can be tuned to a specific frequency to coincide with a selected molecular transition. There are several types of devices with this capability that are now available; their properties are summarized in Table 1. Of these devices, only the dye solution and the parametric oscillator are truly continuously tunable. Molecular-gas lasers oscillate only at discrete frequencies. Spin-flip and diode lasers can emit only within modes characteristic of the amplifying crystal, with gaps between successive modes. The spin-flip laser is tuned by varying the magnetic field applied to the crystal, while the diode laser is tuned by varying the current through the crystal. In the latter case, the crystal must be maintained close to liquid-helium temperature. This requirement presents a serious limitation to widespread deployment of any system based on such a device.

Perhaps the most versatile tunable system is the dye solution laser. The bandwidth and center wavelength of this laser can be varied at will throughout the optical components in the laser cavity. In the pulsed mode of operation, which is the most suitable for atmospheric monitoring, the dye can deliver

Table 1
Sources of tunable optical radiation.

Device	Wavelength range (μm)	Power	Comments	Reference
Molecular laser	2.64–3.05 (HF) 5.08–6.66 (CO) 9.13–11.04 (CO_2, N_2O)	1 W (c.w.), $> 10^6$ W (pulsed) 1 W (c.w.), 10^3 W (pulsed) 6×10^4 W (c.w.), $> 10^7$ W (pulsed)	All molecular lasers oscillate at discrete lines in the spectrum of the gas. Thus, resonant absorption of a line by another gas is a matter of the chance overlap of the two transitions	[17]
Spin-flip Raman	5–6 (CO pump) 9–14 (CO_2 pump)	up to 10% of pump laser power	Magnetic field required for operation	[18]
Tunable diode	5–6 ($PbS_{1-x}Se_x$) 8–16 ($Pb_{1-x}Sn_xTe$)	$< 10^{-4}$ W (c.w.)	Liquid helium required for operation	[19]
Dye solution	0.35–1.0	10^5 W (flash-pumped) 10^8 W (laser-pumped)	Continuously tunable	[20]
Parametric oscillator	0.65–3.0	$\sim 10^3$ W at 75 pulses/sec	Pumped by frequency-doubled Nd:YAG laser	[21]

several tenths of a joule of optical energy in about a microsecond (if flash-pumped) or 10 nanoseconds (if pumped by a Q-switched ruby or Nd:YAG laser). Flash-pumped dye lasers are particularly easy to build [22]. The broad homogeneous gain curve of a dye permits one to suppress oscillation at certain wavelengths by incorporating an absorber of those wavelengths in the laser cavity [23]. This would allow, for example, selective detection of a single component in a mixture of different species in which the spectra overlapped each other.

The optical parametric oscillator, which operates by mixing fundamental and second-harmonic radiation from a Nd:YAG laser in a temperature-tuned nonlinear crystal such as lithium niobate, is still a complex and expensive device. However, it may turn out to be the only good source of tunable radiation in the 3-μm range, which is important for C-H vibrations in hydrocarbons.

6   A Specific Proposal: Monitoring of Atmospheric NO_2

The major visible constituent of photochemical smog is NO_2, a toxic brown gas which is the thermodynamically stable form of nitrogen oxides

('NO$_x$') in the atmosphere. Its visible absorption spectrum is shown in Figure 3. The distribution of this pollutant can easily be determined by long-path resonant absorption monitoring.

Using a flashlamp-pumped coumarin or umbelliferone dye laser, an output of 0.1 joule at 4,200–4,400 Å can easily be obtained, in the form of a 1-microsecond pulse at 100 kW peak power. This corresponds to a pulse of 2.5×10^{17} photons. The beam would be sent through the polluted atmosphere to retroreflectors at a distance of 2 km. With suitable optics, the beam divergence can be made as small as 0.1 milliradian, so that the initially 1-cm² area beam expands to a diameter of 20 cm after travelling 2 km. The beam is reflected by a 10×10 cm mosaic of retroreflectors, similar to that used in obtaining laser-beam echoes from the moon in the Apollo 11 experiment. The retroreflectors should not produce any additional beam divergence, which is 0.1 milliradian, or 18 arc seconds. Since retroreflectors can be fabricated to an order of magnitude higher collimation than this, this presents no problem. If the beam returns to a 5-cm diameter phototube, the net geometrical loss after a 4-km round trip will be a factor of 3×10^3.

A significant additional loss will be that caused by Rayleigh and Mie scattering, especially in the hazy atmosphere often accompanying high NO$_2$ smog concentrations. The attenuation coefficient has been estimated at 1 (km)$^{-1}$ [2], leading to a loss of an additional factor of 50 over a 4-km path. Thus, the net return signal we can expect is on the order of 10^{12} photons/pulse – nearly ten orders of magnitude better than in the Raman or fluorescence backscattering schemes.

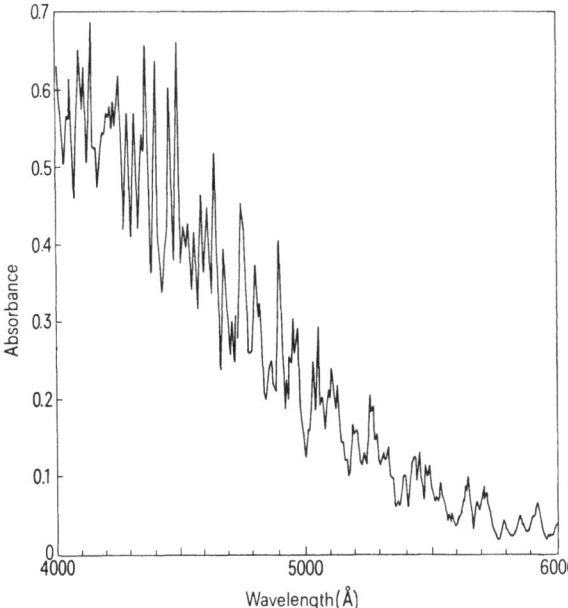

Figure 3
Absorption spectrum of
6 Torr NO$_2$ (pressurized
with 1 atmosphere of air)
in a 12.7-cm cell,
between 4,000 and 6,000 Å

Measurement of the pollutant concentration depends on the variation of atmospheric transmission coefficient with wavelength. Figure 3 shows strong absorption peaks of NO_2 near 4,350 Å, 4,390 Å, 4,460 Å and 4,480 Å, with effective absorption coefficients approximating 1.0 (ppm-km)$^{-1}$. These peaks are separated by absorption minima in which $k(\lambda) \simeq 0.43$ (ppm-km)$^{-1}$. By using (4), we can estimate the differential absorption for various concentrations of NO_2. This is shown in Table 2.

The major interference in these measurements will be diffuse reflections of background sky radiation accepted by the photomultiplier. The spectral radiance of the sky is approximately 0.01 watts cm^{-2} micron^{-1} steradian^{-1} near 4,500 Å [24]. If the field of view of the photomultiplier is restricted to 1°, the spectral bandpass limited to 10 Å around the laser line to be detected, and the photomultiplier properly gated to the 1 microsecond duration of the returning pulse, then we would expect approximately 10^6 photons to strike the photomultiplier. Since this is 10^5 to 10^6 times smaller than the laser photon flux to be measured, this represents a negligible background. At night, of course, this background will be at least a factor of 200 less important. In addition to this background, there will be a component from light elastically scattered by the 150-m long column of air immediately in front of the retro-reflectors during the microsecond that the photomultiplier is turned on to accept the returning laser pulse. But this is estimated to be only about 10^5 photons under the stated conditions, and thus is also negligible.

Actually, the estimated rate of return of 10^{11} photons in a 10^{-6}-second interval is far beyond the linear response region of most photomultipliers. Thus, in addition to gating, passive attenuation of the detected light by several orders of magnitude would probably be required. Background radiation would, of course, be attenuated by the same factor as the reflected laser pulse.

A fairly straightforward variation of the basic resonant absorption technique may be able to provide depth resolution of pollutant concentrations as well. This approach would take advantage of the wavelength variation of the atmospheric transmission coefficients T_1 and T_2 in (1) for backscattered radiation. A laser-pumped dye laser, delivering 10^{17} photons in 10^{-8} sec,

Table 2
Absorption of laser pulses by atmospheric NO_2.

Concentration (ppm)	Percent absorption in 4-km path at absorption peak	at absorption minimum	No. of photons returned
1	98	82	~10^{11}
0.5	88	58	3×10^{11}
0.1	35	16	8×10^{11}
0.03	12	5	9×10^{11}
0.001	0.4	0.2	~10^{12}

would be used. If the elastic scattering coefficient is $1.0\,(km)^{-1}$ [2], then 3×10^{14} photons will be scattered out of the 3-m column of polluted air that corresponds to the 'length' of a 10^{-8}-second light pulse. A 5-cm diameter phototube will intercept one part in 10^{10} of the light scattered at a range of 1 km, so that approximately 3×10^4 photons could be collected in the 10^{-8}-second detection interval. This is of the order of the daytime background light flux, and so should be detectable. This measurement would give a rough estimate of the integrated pollutant density,

$$\bar{\varrho}(R) = \int_{0\to R\to 0} \varrho(r,\,\theta)\mathrm{d}r, \qquad (5)$$

where R is the range from which the backscattering is received. These low-accuracy measurements, combined with the high-accuracy determination from reflected beams at fixed distances, should be able to provide a fairly detailed map of atmospheric NO_2 concentrations.

Other systems. There are a number of other species that could be monitored by the resonant absorption technique. SO_2 has an absorption in the 3,500 to 3,900 Å region, but the absorption coefficient is only $0.0004\,(ppm\text{-}km)^{-1}$. A sufficiently large absorbance is found at $\lambda \sim 3,100$ Å, which can be probed with dyes pumped by a quadrupled neodynium laser. A prototype of a similar system for monitoring ammonia and ethylene, using several lines of the CO_2 infrared laser, has been tested by the General Electric Company [25].

Atomic species may be particularly easy to detect by long-path absorption; since their absorption coefficients can be on the order of 10^3 larger than in the molecular case, the ultimate sensitivity for atomic species may be in the parts-per-billion range, rather than the parts-per-million range as for molecules.

7   Summary and Conclusions

Of the three monitoring methods considered, namely, Raman backscattering, resonance fluorescence backscattering, and long-path absorption, the latter seems to be the most promising for atmospheric surveillance. Even for depth resolution from a fixed monitoring station, measurement of transmission variation from elastically backscattered laser light is competitive with Raman or fluorescence detection. In the specific case of NO_2, sensitivity limits are far smaller for the absorption method than for the other two methods.

What is needed in this area is a systematic research program to evaluate, test, and deploy these monitoring schemes. Unfortunately, support for such research does not seem to be readily forthcoming. It is a sad commentary on our national priorities that, thanks to well-funded research and development programs, laser guidance systems can be used to direct bomb to targets with better than 2-m accuracy [26], while development of a parallel technology for the improvement of our environment and quality of life is stymied by lack of both research funds and official interest.

Once the necessary research can be carried out, it is not hard to envision the widespread deployment of a variety of monitoring programs for specific purposes. Atmospheric monitoring stations, located on tall buildings in a metropolitan area, would be used to maintain a constant watch over ambient pollution levels. These data could be used to test the results of control strategies, or, in the event of failure, to alert the public when a danger level of some pollutant had been reached. Local monitoring stations, perhaps based on tunable infrared diode lasers, could be used to detect specific emission levels in closed environments such as furnace exhausts, or carbon monoxide levels in vehicular tunnels. Whenever any of these indicators reached a danger level, local authorities would have the obligation to curtail activities, by shutting down plants, limiting traffic through the tunnel, prohibiting use of automobiles, or other steps. Perhaps only some drastic restrictive measures such as these will finally provide the public incentives for widespread development of nonpolluting forms of manufacturing and transportation.

Acknowledgments

This work was supported by the New England Consortium on Environmental Pollution, through a subcontract with the Lowell Technical Institute Research Foundation under Contract No. 68-02-0314 from the Environmental Protection Agency.

References

[1] G. B. Morgan, G. Ozolins and E. C. Tabor, Science *170*, 289 (1970).
[2] H. Kildal and R. L. Byer, *Comparison of Laser Methods for the Remote Detection of Atmospheric Pollutants*, Report No. AD-730770 (National Technical Information Service, Springfield, Virginia, 22151).
[3] M. McClintock, *Air and Water Pollution* (Eds. W. E. Britton, R. West and R. Williams; Colorado Associated Universities Press, Boulder 1972), p. 477.
[4] T. Kobayasi and H. Inaba, Proc. I.E.E.E. *58*, 1568 (1970).
[5] T. Hirschfeld and S. Klainer, Opt. Spectra, p. 64 (July/August 1970).
[6] T. Kobayasi and H. Inaba, Appl. Phys. Lett. *17*, 139 (1970).
[7] J. I. Steinfeld, Accts. Chem. Res. *3*, 313 (1970).
[8] For reviews, see R. G. Gordon, W. Klemperer and J. I. Steinfeld, Ann. Rev. Phys. Chem. *19*, 215 (1968); J. I. Steinfeld, *MTP International Review of Science*, Phys. Chem. [I] *9*, chap. 8 (Ed. J. C. Polanyi; University Park Press, Baltimore 1972).
[9] *Handbook of Military Infrared Technology* (Ed. W. L. Wolfe; US Government Printing Office, Washington, D.C., 1965), p. 462–500.
[10] R. T. Menzies, Appl. Opt. *10*, 1532 (1971).
[11] E. D. Hinkley and P. L. Kelley, Science *171*, 635 (1971).
[12] P. L. Kelley, E. D. Hinkley and A. R. Calawa, Proc. 7th Int. Quantum Electronics Conf. (Montreal, Québec, May 1972), p. 87; E. D. Hinkley, A. R. Calawa, P. L. Kelley and S. A. Clough, J. Appl. Phys., to be published.
[13] K. W. Nill, F. A. Blum, A. R. Calawa and T. C. Harmon, Proc. 7th Int. Quantum Electronics Conf. (Montreal, Québec, May 1972) p. 88.
[14] E. D. Hinkley (private communication) has succeeded in transmitting a diode laser beam over a range of ~ 100 meters.

[15] L. B. KREUZER and C. K. N. PATEL, Science *173*, 45 (1971).
[16] A. KALDOR, W. B. OLSON and A. G. MAKI, Science *176*, 508 (1972).
[17] *CRC Handbook of Lasers (and) Optical Technology* (Chemical Rubber Co., Cleveland, Ohio 1971), p. 298–349.
[18] C. K. N. PATEL, Phys. Rev. [B]3, 1279 (1971).
[19] E. D. HINKLEY, Appl. Phys. Lett. *16*, 351 (1970).
[20] B. R. SNAVELY, Proc. I.E.E.E. *57*, 1374 (1969).
[21] R. W. WALLACE and S. E. HARRIS, *Extending the Tunability Spectrum*, Laser Focus (November 1970).
[22] C. L. STONG, Scient. Am. *222*, 116 (February 1970).
[23] R. A. KELLER, E. F. ZARLENSKI and N. C. PETERSON, Bull. Am. Phys. Soc. *16*, 442 (1971).
[24] W. K. PRATT, *Laser Communication Systems* (John Wiley and Sons, New York 1969); also see [2], p. 76.
[25] L. R. SNOWMAN and R. J. GILLMEISTER, Opt. Spectra, p. 30 (June 1972).
[26] Newsweek, p. 53 (June 1972).

X-Ray Fluorescence – an International Calibration Method

by Miss J. M. DESHPANDE
Central Public Health Engineering Research Institute, Nagpur, India

Abstract

To get an idea of global pollution the data collected in different countries have to be compared. Comparison is only valid if all the different methods are calibrated against some standard method. The standard method should be simple and subject to minimum error. The estimation of sulfur dioxide and particulate matter by X-ray fluorescence seems to fulfill this condition, hence it may be adopted as an International Calibration Method.

1 Introduction

Mathematical models may predict the theoretical amount of air pollution to be expected in a particular area. But only analytical methods can test its validity to give a true picture of the magnitude of the air pollution problem. Since air pollutants are present in microquantities in a dynamic phase subject to change of pressure, temperature and relative humidity, it is difficult to estimate individual pollutants accurately without incurring positive or negative interference from other pollutants present simultaneously in the atmosphere. To estimate the pollutants in question with accuracy and precision, sometimes certain inhibitors have to be added to eliminate the influence of interferences [1]. Thus it makes the method complicated and difficult to handle and gives rise to additional experimental errors. This is why new procedures are being developed which incorporate different physical properties such as colorimetric, coulimetric, chemiluminance, infrared or X-ray fluorescence in order to be able to estimate air pollutants with precision and accuracy.

2 The X-Ray Fluorescence Method

The phenomenon of X-ray fluorescence (XRF) which generally is a disturbing factor in diffraction experiments, has been used to estimate airborne particles by GRENNFELT et al. [2]. The fluorescent effect is produced by bombarding the elements with high energy X-rays; the primary radiation gives rise to secondary radiation, which can be measured in an X-ray spectrometer. The relationship between the concentration of an element and the intensity of

4   Views on Air Pollution Problems in Developing Countries

The advantages in adopting the X-ray fluorescence method as an international calibration method, especially the simplicity in procedure and its low cost, could enable developing countries to use it for their urgently needed air pollution monitoring systems.

With a rapid growth of industrialisation and urbanisation in developing countries, local concentrations of air pollution are being encountered in more and more places. To monitor air pollution and to keep it under control involves the development of a technology which exists only partially even in the most highly industrialized countries. Unfortunately developing countries cannot hope to acquire such an advanced technology in the near future. In developing countries the prime objective of the governmental policy is to increase industrialization in order to create jobs and raise the standard of living. Too stringent antipollution measures might hamper this development and might thus not be favored by a growth-oriented government. Only inexpensive antipollution devices would allow both a development and a protection of the environment. Until such devices are on the market, the developing countries will continue to ignore pollution control in favor of continued industrial development.

There is another basic problem: The question about the validity of present air quality criteria for developing countries. Until now most work in the field of air pollution has been carried out in highly developed countries which happen to have temperate climate. Most of the developing countries are tropical. In the tropics the dust concentration is 4–5 times that found in temperate climates. Particulates may play a more significant role when other pollutants are present to hasten the chemical reaction. Standards which are reasonable for a temperate climate may not be adequate in a tropical climate. Consequently other air quality criteria have to be developed for the Third World.

We thus have this dilemma: Since the developing countries are still in the process of growing they could learn from the environmental mistakes and pollution problems of the industrialized nations. But due to their financial situation, they cannot afford the existing expensive technology. In other words: what they need are simple, cheap and easy devices to control and monitor environmental pollution.

Acknowledgment

The authoress is gratefull to Professor CYRILL BROSSET, Swedish Water and Air Pollution Research Laboratory, for his guidance.

References

[1]   HERBERT C. MCKEE, R. E. CHILDERS and OSCAR SAENZ, JR., *Collaborative Study of Reference Method for Determination of Sulfur Dioxide in the Atmosphere (Pororos-aniline Method)*, Contract CPA 70-40 SWRI Project 21-2811. Prepared for the office

of measurement standardization, Division of Chemistry and Physics, National Environment Research Center, Environment Protection Agency (September 1971).

[2]   P. Grennfelt, Å. Åkerström and C. Brosset, *Determination of Filter Collected Airborne Matter by X-Ray Fluorescence*, in: *Atmospheric Environment*, Vol. 5 (Pergamon Press, 1971), p. 1–6.

[3]   Thomas R. Dittrich and C. Richard Cothern, *Analysis of Trace Metal Particulates in Atmospheric Samples Using X-Ray Fluorescence*, J. Air Pollut. Control Ass. *21*, No. 11, 716–719 (1971).

[4]   M. Bonnevie-Svendsen and A. Follo, *Comparision with Water Soluble Sulphate in Sample Filters*, NILU Report, IFA CH98, Kjeller Norway (March 1972).

[5]   P. Grennfelt and Cyrill Brosset, *Determination of Sulfur Dioxide by Means of Selective Absorption on Impregnated Filters and Analysis by X-ray Fluorescence*, Swedish Water and Air Pollution Research Laboratory, Publication report B122, Gothenburg (June 1972).

[6]   P. Urone, J. B. Evans and C. M. Noyes, *Tracer Techniques in Sulfur Dioxide Colorimetric and Conductiometric Methods*, Anal. Chem. *37*, 1104 (1965).

Nitric Oxide Formation in Industrial Combustors

by D. Thompson
Department of Physical Chemistry, University of Leeds, England

Abstract

Fuel-lean combustion of liquid and gaseous hydrocarbons is a major source of power, but consequent emission of hydrocarbons and nitric oxide leads to the formation of photochemical smogs. Reduced nitric oxide emission by combustors would greatly reduce the incidence of the condition.

Experimental investigation of fuel-lean premixed methane-air combustion gave agreement between observed nitric oxide formation rates and those predicted by the ZELDOVICH *mechanism, when non-equilibration of atomic and molecular oxygen concentrations was taken into account.*

Consideration of this and other reported characteristics of nitric oxide formation in flames leads to preference for stabilized premixed combustion *of volatile fuel. It is concluded that suitable stationary units could be introduced; in the motive power field preference for an alternative system to the internal combustion engine is indicated and discussed.*

1   Introduction

The increasing use of liquid and gaseous fuels has led to the appearance of a widespread air pollution condition known as photochemical smog, caused by hydrocarbons and nitrogen oxides, and causing the eye and nose irritation characteristic of the condition (due to ozone, acrolein, peroxyacyl nitrates or PAN) and long-term effects on plant life (ozone and PAN) and manufactured goods, especially rubber (ozone) [1].

Typically atmospheric levels of nitrogen oxides and hydrocarbons associated with severe smog episodes are 1 ppm and 4 ppm, respectively [2]. Experimental investigations have shown that there is a wide range of smog formation potential associated with various hydrocarbons; thus propane has very little potential, and aldehydes are highly effective. Both nitrogen oxides and hydrocarbons are necessary smog ingredients – in a system involving only nitrogen oxides, oxygen and sunlight, ozone levels would be much less than those observed in smogs, and the higher concentration is attributed to reactions involving hydrocarbon species. (For a detailed investigation of the photochemical smog in Europe see Formosinho and Cardodo's chapter in this book.) In order to define an approach to elimination of the condition it is necessary to examine the general characteristics of emission and pollution by combustion of these fuels, and the prospects for successful treatment by pollutant precursor removal either before distribution, or at the combustor before, while, or after it is burnt.

Nitric oxide emission is for all practical purposes confined to combustor exhaust gases and the overwhelming majority of this is a result of the combination of atmospheric oxygen and nitrogen. Thus, whilst considerable reduction in smog formation may be obtained by a reduction of the hydrocarbon emission levels, or use of a fuel which is neither a severe smog precursor or a source of precursors during combustion (e.g., propane), the modifications involve widespread technologies or a drastic change in fuel policy. Conversely, reduction of emission of nitrogen oxides from combustors could solve the problem in one move.

Nitric oxide emission levels from *automotive* internal combustion engines vary from 20 to 4,000 ppm, idling and accelerating, respectively. Emissions from *stationary* combustors under normal operating conditions vary from tens to hundreds of ppm. It has been calculated that the present respective contributions to atmospheric nitrogen oxides levels are approximately equal.

Experimental and theoretical investigations have shown that the nitric oxide level emitted by the internal combustion engine is formed within a very short time after the combustion process has been completed, i.e. *before* the power stroke has progressed far. Levels of emission by a cruising automobile can be reduced by 80% by recycling part of the exhaust gases with acceptably small power loss [2], but this level of reduction would not satisfy the requirements of Californian legislation, and consequently catalytic reduction in the exhaust appears the only viable solution.

The characteristics of nitric oxide formation in furnaces are less intractable; the flame is continuous and accessible, and there is a considerably wider range of freedom of design and control over the flame structure than in the i.c. engine. Thus, a simple solution to the problem of nitric oxide formation in them is not inconceivable.

2   Flame Characteristics Associated with Nitrogen Oxides Emission

Before discussion of the characteristics of flames in relation to the formation of nitric oxide, it is necessary to consider why the species is formed at all. This follows from approach to thermodynamic equilibrium concentrations at the elevated temperatures attained immediately after combustion. Any changes in concentration will move towards equilibrium (unless the species concerned is an intermediate in an intense reaction system). The equilibrium composition of the products of fuel-lean hydrocarbon – air combustion at room temperature includes only CO_2, H_2O, N_2 and excess O_2 but at higher temperatures typical of flames other species are present in small concentrations. These include H, OH, O, H_2, CO and NO, and for instance, in the presence of 5% O_2 and 80% N_2 the equilibrium concentrations of NO are 0.2%, 0.7% and 1.4% at 1,800, 2,200 and 2,600 °K.

Available information indicates that the nitric oxide is always present at a concentration less than equilibrium at the maximum flame temperature before this temperature is attained, and it follows that this concentration is the maximum which may be formed in any particular combustor. At some stage in the cooling of the burnt gases the falling rate of decomposition is unable to maintain equilibrium, and much of the nitric oxide is *frozen out* and appears in the exhaust.

Flame structure in relation to the formation of nitric oxide

The complex structure of many industrial flames does not allow direct evaluation of factors controlling nitric oxide formation in them directly. It is necessary to consider a simplified flame.

Basic flame characteristics may be described with reference to the flat premixed stabilized flame. In this, a stream of gas moving at uniform velocity enters a thin zone (flame region) where the majority of the chemical energy is converted to heat. The gas passing out of the other side is hot, and close to the equilibrium composition.

In the flame zone equilibrium has little significance with respect to the concentration of transiently formed highly active species. In relation to nitric oxide formation, it may be noted that the zone is *thin* (i.e. residence time is *short*), and for much of its thickness temperature is well below the maximum attained. Thus, a fast reaction involving species transiently present only in the flame region would need to be involved for the yield of nitric oxide to be significant compared with the yield during the much longer residence time in the post flame gases. Experimental results have been obtained suggesting that such a process may occur in hydrocarbon flames [2, 3].

In the post flame region chemical equilibrium is a significant property of the gas. In all but the most rapidly cooled, or hottest systems, nitric oxide concentration is below equilibrium for a measureable time. Major species are effectively in equilibrium, but the concentrations of O, OH, H, CO and H_2 are not always equilibrated in flames of hydrocarbons at lower temperatures than the industrial flame range, and it is possible that they are sometimes above equilibrium in the latter. Their persistence is due to the involvement of slow *third body* reactions in their removal.

Experimental investigations of propane-air flames [4] have indicated that the rate of nitric oxide formation in the post flame region may exceed that predicted by 'hot-air' kinetics (assumption that all species but NO are in equilibrium, and only nitrogen and oxygen are involved in the chemical reactions). Thus, it is necessary to know whether this effect is due to the more general non-equilibration referred to above, or to use of an incomplete kinetic model for predictions in other respects.

If this fundamental information about the formation of nitric oxide in the flame and post flame regions could be obtained, we would be in a position to specify the possibilities for the design of a low nitric oxide emission level combustor to satisfy each of the technical applications.

3   Experimental Investigation

The design of the combustor used in this investigation was adopted as a compromise between the characteristics of an industrial burner, i.e. combustible and throughput, and those of an experimental flame with respect to spatial resolution. Reported dependence of nitric oxide formation on the presence of oscillatory combustion conditions was investigated concurrently; the adopted system was resonant and designed to allow variation of the frequency and amplitude of oscillation.

The details of construction and operation have been reported elsewhere [5–7]. Essentially, the burner was a variable length 2.5 cm diameter tube with a multiport flameholder mounted a variable distance from the entrance (Fig. 1). Combustion product concentrations were measured by analysis of low pressure (15 to 20 mm Hg) batch samples obtained by quartz microprobe; nitric oxide was determined by a modification of the Saltzman technique, [8] and the remaining species were determined by mass spectroscopy.

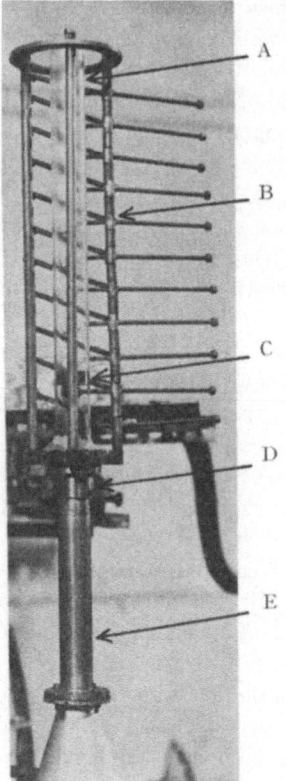

A	Upper burner tube (quartz)
B	Sampling port cover pivot assembly
C	Flameholder
D	Upper burner tube support
E	Lower burner tube

Figure 1
Experimental combustor design. Length 15–43 cm (variable), diameter 2.5 cm.

Combustion conditions

An air flowrate of 98 litre per minute and methane concentrations of 7.3 to 9.2% were used as the most satisfactory compromise between the requirements of spatial profile resolution, restriction of random noise, and similarity of throughput to that of industrial combustors.

The flame appeared as a set of identical anchoring flamelets, each almost cylindrical, and sharply defined, anchored above each port in the flameholder. 2–3 cm beyond the flameholder these merged into a homogeneous blue volume of gas occupying the whole cross section of the tube, which faded with distance downstream, so that with even the richest mixtures investigated it was invisible except in subdued light after 5 to 8 cm. Localized sodium chloride seeding indicated that little mixing occurred, since the resulting zone of yellow gas did not broaden appreciably with distance downstream.

The burner was normally resonant (500–1,250 Hz, 1 psi double amplitude); quiescent combustion was obtained by slight modification. No effect on the nitric oxide yield due to oscillatory combustion was observed, and it was concluded [5–7] that a significant change in yield observed by others [9] was due to gross systematic changes rather than any local effect.

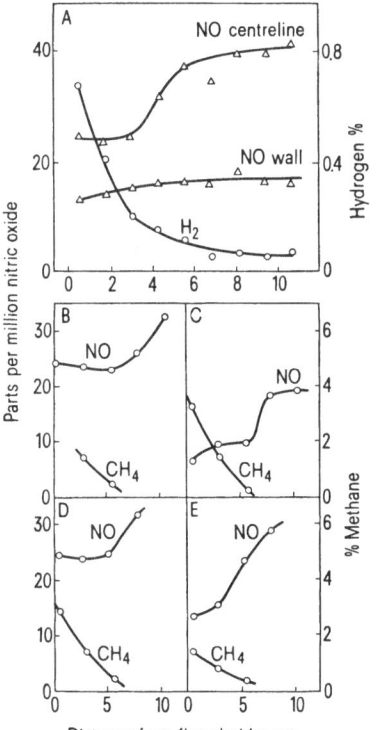

Figure 2

A: Typical nitric oxide and hydrogen concentration profiles. B–E: Nitric oxide profiles in the homogeneous combustion zone.

Experimental concentrations

Typical nitric oxide and hydrogen concentration profiles are shown in Figure 2a. The general form of the nitric oxide profile at the centreline is representative of the results for $CH_4 > 7.3\%$; it was formed immediately above the flameholder (up to 25 ppm), and in cases where more than one result was obtained in the homogeneous combustion zone, the level was reasonably constant (Fig. 2B–2E). A rapid rise in concentration of up to 18 ppm 1 or 2 msec after the majority of the methane had been burnt then occurred when the combustible contained more than 7.3% CH_4. Maximum nitric oxide concentration observed at the centreline increased with fuel concentration and rising flame temperature.

Post flame hydrogen concentrations were always greater than the equilibrium concentration predicted from $K(H_2 + {}^1/_2 O_2 = H_2O)$, by a factor of from 2.7 to 12.

The presence of stabilising flamelets prevented unambiguous descriptate of early nitric oxide formation, and comparison of the observed formation rion with theoretical predictions was confined to the post flame region ($CH_4 < 0.5\%$).

Reaction mechanism

The formation of nitric oxide in the present range of interest ($T > 1,800°K$, $[O_2] < 5\%$) is dominated by the ZELDOVICH *chain mechanism* [10]. This is the reaction of oxygen atoms (commonly assumed to be equilibrated with molecular oxygen) with nitrogen

$$N_2 + O \rightleftharpoons NO + N. \tag{1}$$

This is followed by

$$N + O_2 \rightleftharpoons NO + O. \tag{2}$$

Combining $k_1 = 1.36 \times 10^{11} \exp (75.4/RT)$ mole^{-1} litre sec^{-1} and $K({}^1/_2 O_2 \rightleftharpoons O)$ $= 5 \times 10^{1.5} \exp (59.3/RT)$ mole$^{-1/2}$ litre$^{1/2}$, the NO formation rate ($[NO] \ll [NO]_{eq}$) is

$$\frac{d[NO]}{dt} = 4.31 \times 10^{13} [O_2]^{1/2}[N_2] \exp. (-134.7/RT) \text{ mole litre}^{-1} \text{sec}^{-1}. \tag{3}$$

A second mechanism, the *bimolecular mechanism*, which involves the presence of steady state nitrous oxide and oxygen atom concentrations, which react to give nitric oxide

$$N_2O + O \rightleftharpoons 2NO \tag{4}$$

is more than an order of magnitude slower than the ZELDOVICH chain mechanism within the stated range; this has been shown to be true even under the partial equilibrium conditions discussed below [6].

Non-equilibrium conditions

The experimental results obtained here have indicated that $[H_2]$ exceeded equilibrium. This does not occur in isolation; it is considered [13] that in post flame gases the fast reactions

$$H + O_2 \rightleftharpoons OH + O \tag{5}$$
$$H + OH \rightleftharpoons O + H_2 \tag{6}$$
$$H + H_2O \rightleftharpoons OH + H_2 \tag{7}$$

are maintained close to internal equilibrium (*balanced*).

Under lean flame conditions $[O_2]$, $[H_2O] \gg [O]$, $[OH]$, $[H]$, $[H_2]$ so that they may be considered constants in the equilibrium expressions for each reaction, written for overall, and partial equilibrium. By combining (5) and (7), (6) and (7), and finally substituting these conditions in any one equilibrium expression, we obtain

$$\frac{[O]}{[O]_{eq}} = \frac{[H_2]}{[H_2]_{eq}} = \frac{[H]^{2/3}}{[H]_{eq}} = \frac{[OH]^2}{[OH]_{eq}} \tag{8}$$

where $[\ \ ]_{eq}$ refers to overall equilibrium.

Thus, since $[O]_{eq}$, $[H]_{eq}$ and $[OH]_{eq}$ may be calculated, an estimate of the non-equilibrium concentration of any of these species may be obtained.

The rate equation of the chain mechanism is based on the rate determining step 1, which is dependent on oxygen atom concentration; the dependence is expressed by $K[O_2]^{1/2}$, and to obtain the true rate, (3) must be multiplied by $[O]/[O]_{eq}$.

Application of kinetic data to experimental results

Predicted increments in nitric oxide concentration in the post flame region were calculated by substitution of mean $[O_2]$, $[N_2]$, $[H_2]/[H_2]_{eq}$ and temperature for adjacent sampling stations, and the residence time between them, Δt, in the equation

$$[NO] = 4.31 \times 10^{13} \frac{[H_2]\,[O_2]^{1/2}\,[N_2]}{[H_2]eq} \Delta t \times \exp\left(-134.7/RT\right). \tag{9}$$

$$[\text{mole litre}^{-1}\ \text{sec}^{-1}]$$

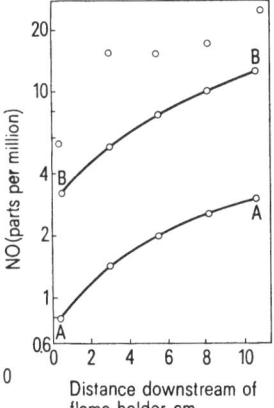

Figure 3

Predicted increments in nitric oxide concentrations using (A) the equilibrium rate equation and (B) the modified rate equation compared to observed values.

The post flame increment for each sampling station was then calculated as the sum of each post flame sampling interval increment up to that point.

The mean ratio of observed values to those predicted by the assumption that [O] and [O_2] were in equilibrium was an order of magnitude; use of the modified rate equation gave considerably improved agreement between theoretical prediction (B) and observation (Fig. 3). The majority of the results agreed within a factor of two (the error limits for the rate constant k_1 suggested by BAULCH et al. [11]), and 87% of the results agreed within a factor of 3. No significant variation of the ratio with distance downstream was noted. The aggregate observed yield/predicted yield for all experiments was 1.7, i.e. the results indicated an experimental formation rate 70% higher than that predicted theoretically, an entirely satisfactory agreement.

4   Discussion

In more general terms, it may be noted that the results indicate that in the post flame region close to the flame good agreement has been obtained between experimental and predicted nitric oxide formation rates, when the predicted rate is obtained by the used of reactions which are flame related only in the sense that the persistence of excess radicals must be taken into account. The reduced rate in the region further downstream is in agreement with these findings, because the high activation energy reaction considered to be responsible for the nitric oxide formation near the flame would be expected to be highly sensitive to decreasing temperature.

In order to give full consideration of the implications of the results in the practical field of smog reduction, it is necessary to take note of the work of WILLIAMS, SAROFIM and LAMBERT [14]. These authors introduced a heat exchanger into the post flame gases of a premixed propane-air flame supported on a multiport flameholder with smaller ports than those of the systems discussed here. It was found that the formation of nitric oxide ceased completely when the burnt gases entered the heat exchanger. No nitric oxide formation was observed in the early flame region, and it may therefore be assumed that the nitric oxide observed in this region in the present work was formed in the high temperatures zones around the stabilising flamelets by the ZELDOVICH mechanism, whereas such zones would be expected to be much smaller in the system investigated by WILLIAMS et al.

5   Application to Air Pollution Reduction

The experimental work has shown that it is possible to offer criteria for the design of a combustor with an acceptably high throughput and highly acceptable from the viewpoint of air pollution reduction. The criteria in their most general form are that the flame (premixed) should have a structure such that the high temperature zones are sufficiently well separated from the flame itself that physical access may be gained to them without interference with the flame processes; cooling should be rapid in these zones.

In application of these findings to practical systems, the basic question asked is: how applicable is the method to existing systems, without radical change in design or fuel?

The commonly available fuels suitable for use in the type of burner proposed are methane (i.e. natural gas), propane, butane and towns gas; gasoline could be used in such a system, since it is readily vapourized; kerosine may also be vapourized, but a well designed heater is required. Heavy fuel oils crack before vapourising, and are not suitable for premixed combustion.

The characteristisc of power stations and stationary combustors in general are compatible with the basic requirements, since a solid heat exchanger is used; the heat transfer efficiency of the exchanger near the flame need only be modest, and a robust design with satisfactory service life under the stringent conditions demanded is feasible. The major limitation at present is probably the extensive use of heavy fuel oil. The burner type recommended here requires a premixed flame, and as heavy oils cannot be vapourized, they are burnt as a droplet spray. The characteristics of such flames are not sufficiently well-known for any predictions about nitric oxide formation in them to be made, and this field is probably the most demanding of attention in future research.

Much of the fuel used in the automotive field is volatile gasoline, and it has even been suggested that the use of propane should be made compulsory, in the interests of reduced hydrocarbon emission, so that there is no criticism on the basis of fuel type requirement. The discussion instead must centre on comparison of the pollutant reduction potential of competitive power units.

At present the field is dominated by the internal combustion engine, in which it is completely impossible to utilize the criteria proposed here. The system is a very severe source of nitric oxide and hydrocarbons; measures taken to reduce its contribution to smogs have included the use of afterburners to burn off hydrocarbons, and future introduction of catalytic beds into the exhaust system appears the only useable means of removing nitric oxide to the degree required by future Californian legislation. Unfortunately the beds are expected to be subject to deterioration and this would require regular renewal of the catalyst, which may be difficult to enforce.

Feasible alternatives to the internal combustion engine must have good prospects, on the basis of present knowledge, for development into a generally acceptable power unit with respect to power/weight ratio, throttle response, refuelling range and operating cost, including servicing. The only alternatives which appear to be at a sufficiently advanced stage of development for consideration are the gas turbine, stored electrical systems and the steam engine. The gas turbine has recently been marketed in commercial vehicles, and electrical 'town cars' and coaches are available, in addition to the long established use of local delivery vehicles powered by this means. The steam engine was used in the early days of the motor car, and for a period was equally, if not more successful than the internal combustion engine. In more recent times its use has been limited to a few isolated vehicles, primarily private conversions of mass production internal combustion engine based vehicles.

The gas turbine is a far less severe source of nitric oxide than the internal combustion engine; the nitric oxide emission of one commercial model has been reported to be comparable to an equivalent internal combustion engine with a 50 ppm nitric oxide exhaust concentration [15] (the level from an aviation engine is higher). The potential for reduction of this level in the exhaust gases is minimal, since the gas from the combustion chamber is highly diluted, but there is some possibility that the level could be reduced by careful adjustment of dilution around the combustion chamber. The major difficulty in the widespread introduction of this power unit as an alternative system for the automobile centres on problems of producing a sufficiently small unit.

The electric car pollution problem regresses to that of the generating station; its use is limited by a combination of range and power – an increase in either would result in a considerable increase in the cost of the storage system.

The steam engine has the two disadvantages of immobility during warm up and slow response on moving off from a halt; the first may be substantially improved by the use of a 'flash' boiler, and the latter (due to the absence of a clutch – i.e. no tickover, and consequent cooling of the cylinders) could be corrected if one were introduced. It is, of course, ideally suited to the combustor design criteria proposed [14].

Comparison of the apparent effects of the results described on the pollution reduction potential of alternative motive power sources is thus considered to suggest that there is a somewhat strengthened case for electric transport, and a much strengthened case for steam. The case for the gas turbine is vaguely strengthened, but no improvement in the status of the internal combustion engine is indicated.

There are thus good grounds for considering that a policy of steam as a general source, or as a substitute in the private sector, where the gas turbine and stored electric vehicle are unsuitable at present could lead to satisfaction of the overall requirements for reduced pollutant emission.

There is a need for much improved understanding of the process of nitric oxide formation in diffusion flames. Subject to indications that emissions from such flames can be reduced to the degree which has been noted for the premixed flame, the apparent preferred policy would be the adoption of general gas or volatile fuel combustion, in systems designed on the basis described.

References

[1] A. J. HAAGEN-SMIT, in: *Air Pollution*, 1st ed. (Ed. Stern; Academic Press, New York 1962), p. 41.
[2] J. BAGG, in: *Air Pollution Control* (Ed. Strauss; W. Wiley, Interscience, New York 1971), p. 35.

[3] C. P. FENIMORE, 13th Int. Symp. Combust. (The Combustion Institute, 1971), p.373.

[4] J. M. SINGER, E. B. COOK, M. E. HARRIS, V. R. ROWE and J. GRUMER, US Bureau of Mines Rept. 6958, presented at Symp. Combust. Reactions of Fossil Fuels, Div. Fuel Chem., Am. Chem. Soc. (September 1966).

[5] D. THOMPSON, Ph. D. Thesis (Sheffield 1971).

[6] D. THOMPSON, T. D. BROWN and J. M. BEÉR, Combust. Flame *19*, 69 (1972).

[7] D. THOMPSON, T. D. BROWN and J. M. BEÉR, presented at 14th Intl. Symp. on Combustion; Penn. State Univ. August 1972.

[8] *Air Pollution*, 2nd ed. (Ed. Stern; Academic Press, New York 1962), p.80.

[9] G. N. RICHTER, H. C. WIESE and B. H. SAGE, Combust. Flame *6*, 1 (1962).

[10] J. ZELDOVICH, Acta phys.-chim. URSS *21/4*, 577 (1946).

[11] D. L. BAULCH, D. D. DRYSDALE, D. G. HORNE and A. C. LLOYD, *Critical Evaluation of Rate Data for Homogeneous Gas Phase Reactions of Interest in High Temperature Systems*, No. 4.

[12] F. KAUFMAN and L. J. DECKER, 7th Int. Symp. Combust. (Butterworths, London 1959), p. 57.

[13] C. P. FENIMORE, *Chemistry in Premixed Flames*, Int. Encycl. phys. Chemistry chemical Phys. Topic 19, Vol. 5 (Pergamon, New York 1964).

[14] G. C. WILLIAMS, A. F. SAROFIM and N. LAMBERT, to appear in Proc. General Motors Symp. on Emissions from Continuous Combust. Sources, to be published by Plenum Press.

[15] D. S. SMITH, R. F. SAWYER and E. S. STARKMAN, J. Air Pollut. Control Ass. *18/1*, 31 (1968).

Eco-Management

'Good management is good planning!' This saying must also be true for ecological management. A new technology is only successful if it is managed effectively. In general, two fields of management may be distinguished: the short-range or day-to-day planning process and the long-range planning. Very little theoretical work has so far been done to improve the first type of management operation system: the day-to-day management of air polution control is often one of the weakest links in today's environmental battle.

We apply still mostly a trial-and-error method. How long can we afford to play the game the old-fashioned way? New approaches and ideas on how to apply various emission standards effectively, for instance, are urgently needed. A new and interesting concept for a dynamic emission control is suggested by Nitin Patel of India, who is presently at the Massachusetts Institute of Technology. It is based on operations research techniques and may have much to offer controlling large sources of pollution, such as power stations or industrial complexes.

The importance of the second management aspect has been stressed in the introduction of this book. We must develop models to check present managerial policies on their eventual adverse long-term effects. And we must establish techniques to evaluate and compare different environmental policies as accurately as possible.

In order to achieve an effective comparison of various complex alternatives, the impact and importance of different pollutants must be based on a common reference system. A new method is suggested by STANISLAW CHROSCIEL and MACIEJ NOWICKI of the Warsaw Technical University in Poland. Their 'Pollution Dose Factor' makes it possible to investigate complex pollution situations in a consistent way. In their article on the optimal location of urban and industrial agglomerations they stress the point made at the very beginning of this book: a long-range planning is necessary in order to cope efficiently with the environmental crisis. But a long-range planning model is only as good as the information supplied as input. In order to succeed in the environmental struggle, we thus need a close cooperation between science, engineering and management.

A Markov Process Model for Dynamic Control of Emissions

by Nitin R. Patel
Massachusetts Institute of Technology, Cambridge, Massachusetts, USA

Abstract

A simple model is developed to evaluate the potential of emission control policies that depend upon meteorological forecasts. An optimal control policy is derived for this model and related analytical results are obtained. A more realistic model leading to a linear programming formulation is discussed and illustrated by means of an example.

1 Introduction

One of the basic problems in the implementation of air pollution control programs is the translation of air quality standards into emission standards. *Air quality standards* specify how often various pollutant concentrations can be permitted to occur and are based on pollutant levels which have negligible or no effect on the health of men, animals and plants. In the US, for example, the Environmental Protection Agency specifies that the daily average concentration of sulfur dioxide should not exceed 365 $\mu g/m^3$ more than once a year, and the annual average concentration should be less than 80 $\mu gm/m^3$ [1]. In order to achieve a given air quality standard it is necessary to decide by how much emissions from the various sources will have to be reduced. For example, if a source is emitting 2,000 lbs/day of SO_2 it may be possible to meet the required air quality by reducing emissions to 1,000 lbs/day. The figure of 1,000 lbs/day would then be the emission standard for that source. *Emission standards* thus specify how much pollutant an emitter will be allowed to release into the atmosphere.

The most widely used method of setting emission standards is to fix an upper limit on the quality of emissions that may be released on any day. This kind of standard will be referred to as a *static emission standard*. In this paper, I propose to argue the merits of setting *dynamic emission standards*. These are emission standards which will meet the required air quality standards but which set down different emission levels for different meteorological conditions. By this approach, emission standards can be set so that when atmospheric conditions favor rapid dispersal of pollutants they will permit larger amounts of emission, but when the atmosphere is stagnant, only small amounts of emission will be allowed. The setting of dynamic standards leads to greater efficiency in the utilization of air resources in that given air quality standards can be met at lower cost than static standards, or alternatively, better air quality can be achieved for a given cost. I do not mean to imply that dynamic standards are universally superior to static standards; they are considerably more complex to set and to administer. However, I feel that they have much to offer in controlling large sources of pollution, such as power stations, kraft paper mills and municipal incinerators. For space heating of homes and offices, and for automobiles, static standards are, perhaps, the only feasible means of control.

In the next few sections I indicate how operations research techniques may be used in setting dynamic emission standards and in quantifying the advantages of dynamic over static emission control policies.

2   A Simple Model

In this section, I will describe and analyze a simple model for air pollution control which, though it is quite crude, does bring into focus the principal elements involved in setting dynamic emission standards.

I shall take as starting point a model, suggested by MARCUS [2]. This model, based on a 'box' model for diffusion, relates concentrations of a pollutant on two successive days as follows:

$$C(t) = (C(t-1) + e(t))\, v(t) \tag{1}$$

where $C(t)$, $C(t-1)$ are the concentrations of the pollutant on days t and t-1, respectively,

$e(t)$ is the amount of pollutant emitted on day t per unit volume of the diffusion 'box'.

$v(t)$ is the 'ventilation factor' for day t.

For our simple model we will assume that $v(t)$ can take on values of either 0 or 1 (that is, days either have perfect ventilation or complete stagnation). Further, let us suppose that $\{v(t)\}$ forms a homogenous Markov process such that

$$pr\,(v(t) = 0 \mid v(t-1) = 0) = p_0 \text{ and } pr\,(v(t) = 1 \mid v(t-1) = 0) = 1\text{-}p_0,$$
$$pr\,(v(t) = 1 \mid v(t-1) = 1) = p_1 \text{ and } pr\,(v(t) = 0 \mid v(t-1) = 1) = 1\text{-}p_1.$$

Let us suppose that we wish to set two levels of emission, level 1 corresponds to an emission of 1 unit of pollutant/day and level 2 corresponds to an emission of 2 units of pollutant/day. Let us suppose level 2 corresponds to a cost of \$0/day (no abatement) whereas level 1 corresponds to a cost of \$$d$/day due to abatement (fuel switching, curtailed production value, etc.).

We have a single air quality standard to meet, namely that a given level of concentration C_0 should not be exceeded more often than $100\alpha\%$ of days. We wish, in other words, to set a dynamic standard such that

$$pr\,(\text{a randomly chosen day has concentration} > C_0) \leqslant \alpha. \tag{2}$$

Clearly if we can meet this dynamic standard using level 2 only, then the optimal standard is to specify this level for all days. This would be the uninteresting case where no abatement is ever required. Let us suppose, to avoid triviality, that (2) cannot be satisfied in this way.

We shall determine an optimal dynamic policy of the following type:

Use level 1 on days when concentration reaches or exceeds a level k, otherwise use level 2 as the emission standard.

Now $k < C_0$, otherwise we will not be able to meet requirement (2). We wish to find what should be the optimal k from $k = 0, 2, 4, \ldots C_0$ (assume C_0 is even).

If we select a day at random from a long-time period, the probability that we obtain a day which has concentration > 0 is the probability that $v = 1$ on that day.

From the theory of Markov chains we know that this probability is given by the steady-state probability of being in ventilation state 1. Let π_0, π_1 be the steady-state probabilities of being in ventilation states 0 and 1, respectively. π_0 and π_1 are given by

$$p_0 \pi_0 + (1 - p_1) \pi_1 = \pi_0 \quad \text{and} \quad \pi_0 + \pi_1 = 1.$$

Solving, we obtain $\pi_0 = \dfrac{1 - p_1}{2 - p_0 - p_1}$ and $\pi_1 = \dfrac{1 - p_0}{2 - p_0 - p_1}$.

In other words pr (a randomly selected day belongs to *some* stepped block of the type shown in Fig.1) $= \pi_1$

$$= \frac{1 - p_0}{2 - p_0 - p_1}. \tag{3}$$

Now, pr (random day belongs to a 'block' of base-length b | it belongs to some block)

$$= \frac{b \times pr \text{ (random day belongs to a block of base-length } b)}{\displaystyle\sum_{k=1}^{\infty} k\, pr \text{ (random day belongs to a block of base-length } k)},$$

$$= \frac{b\, p_1^{b-1} (1 - p_1)}{E \text{ (base of blocks)}} = \frac{b\, p_1^{b-1} (1 - p_1)}{1/(1 - p_1)} = \frac{b\, p_1^{b-1} (1 - p_1)^2}{b = 1, 2, 3. \ldots} \tag{4}$$

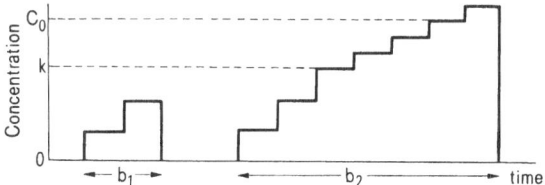

Figure 1
Pollution concentration over time as a stepfunction.

Also,

pr (random day has concentration $> C_0$ | it belongs to a block of base-length b)

$$= \frac{1}{b}\left(b - C_0 + \frac{k}{2}\right) \quad \text{for} \quad b \geqslant C_0 - \frac{k}{2},$$

$$= 0 \qquad \qquad \text{otherwise.} \tag{5}$$

Combining (3), (4) and (5) we obtain

pr (random day has concentration $> C_0$)

$$= \sum_{b=C_0-k/2}^{\infty} \frac{1}{b}\left(b - C_0 + \frac{k}{2}\right) b\, p_1^{b-1}(1-p_1)^2 \frac{(1-p_0)}{(2-p_0-p_1)},$$

$$= \frac{(1-p_0)\, p_1^{C_0-k/2}}{(2-p_0-p_1)}.$$

Clearly to minimize cost we should arrange to make k as large as possible consistent with satisfying requirement (2), i.e. k must be the largest even integer for which

$$\frac{(1-p_0)\, p_1^{C_0-k/2}}{(2-p_0-p_1)} \leqslant \alpha \quad \text{or} \quad \frac{(1-p_0)\, p_1^{C_0}}{\alpha\,(2-p_0-p_1)} \leqslant p_1^{k/2}. \tag{6}$$

Since the log function in monotone increasing

$$\log(1-p_0) + C_0 \log p_1 - \log \alpha - \log(2-p_0-p_1) \leqslant \frac{k}{2} \log p_1$$

so that $k/2$ must be the largest integer $(k \leqslant C_0)$ such that

$$\frac{k}{2} \leqslant \frac{\log \dfrac{(1-p_0)}{(2-p_0-p_1)\,\alpha}}{\log p_1} + C_0$$

or more simply $\dfrac{k}{2} \leqslant \dfrac{\log(\pi_1/\alpha)}{\log p_1} + C_0.$ \hfill (7)

We have thus obtained the optimal dynamic emission policy that will meet air quality standards at lowest cost. (Note that the actual value of d plays no role in the analysis thus far.)

We shall next obtain the average cost per day of the optimal dynamic policy for comparison with the static policy.

pr (Random day has concentration $\geqslant k$ | it belongs to a block of base-length b)

$$= \frac{b - k/2 + 1}{b} \quad \text{for } b > k/2,$$

$$= 0 \qquad\qquad \text{otherwise.}$$

pr (Random day has concentration $\geqslant k$)

$$= \sum_{b = k/2}^{\infty} \left\{ \frac{2b - (k-2)}{2b} \right\} \, b \, p_1^{b-1} (1 - p_1)^2 \, \frac{(1 - p_0)}{(2 - p_0 - p_1)},$$

$$= \pi_1 \, p_1^{k/2 - 1}. \tag{8}$$

For optimal k given by (6),

$$E \text{ (cost)} = \frac{d \, \pi_1}{p_1} \cdot \frac{\pi_1 \, p_1^{C_0}}{\alpha} = \frac{d \pi_1^2 \, p_1^{C_0 - 1}}{\alpha}. \tag{9}$$

Let us now develop for purposes of comparison the cost for a static emissions policy.

Let $1 \leqslant f \leqslant 2$ be the static emission standard per day.

Using analysis similar to that used above we obtain.

pr (Random day has concentration $> C_0$) $= \pi_1 \, p_1^{[C_0/f]}$

where the notation $[x]$ indicates the largest integer $\leqslant x$.

To minimize cost we will choose the largest value of f such that $\pi_1 \, p_1^{[C_0/f]} \leqslant \alpha$

$$\text{or} \quad \left[\frac{C_0}{f} \right] \geqslant \frac{\log \alpha/\pi_1}{\log p_1}. \tag{10}$$

Let $\phi \, (f)$ be the cost ($\$$/day) of restricting the emission per day to a value of f. (Note $\phi \, (2) = 0$ and $\phi \, (1) = d$.)

Then the expected cost for the static policy will be $\phi \, (f^*)$ where f^* is the solution to (10).

Computation of the average costs for dynamic and static emission control strategies for a hypothetical example are shown below.

$$\text{Let } d = 1, \, p_0 = p_1 = \frac{1}{2}, \, \alpha = \frac{1}{256}, \, C_0 = 8.$$

Let $\phi \, (f)$ be linear

Then we find $\pi_0 = \pi_1 = 1/2$.

$$E \text{ (cost dynamic)} = \frac{(1/4) \, (1/2)^{8-1}}{1/256} = \frac{1}{2}.$$

Also,

$$\left[\frac{C_0}{f}\right] \geqslant \log \frac{\dfrac{1/256}{1/2}}{\log\ 1/2} = 7,\ f^* = \frac{8}{7}.$$

$$E\ (\text{cost static}) = 1\left(2 - \frac{8}{7}\right) = \frac{6}{7}.$$

$$\text{Ratio dynamic to static cost} = \frac{7}{12}.$$

3   A More Complex Model

The model discussed in the previous section needs to be expanded to make it more realistic. In this section we will consider a model which is expanded in order to: a) incorporate ventilation states other than total and zero ventilation, b) have air quality standards specified by both a constraint of type (2) and an average annual concentration constraint, c) have more than two levels of emission standards.

Again we use as our starting point relation (1)*. We shall assume that there are n levels of ventilation $v_1, v_2, \ldots, v_n$ $(0 \leqslant v_j \leqslant 1, j = 1, 2, n)$. Let there be q levels of emission to be incorporated in the dynamic standard $e_1, e_2, \ldots e_q$ (with $e_1 < e_2, \ldots < e_q$). Let us also restrict concentrations to $C_1, C_2, \ldots C_m$.

We shall assume that the sequence $\{v(t)\}$ forms a homogenous Markov chain on the n states $v_1, v_2, \ldots v_n$. Let $q(j \mid L) = pr\ \{v(t) = j \mid v(t-1) = L\}$. From relation (1) using discretization we can compute

$$p(i, j \mid k, L, d) = pr\ (C\ (t) = i, v(t) = j \mid C(t-1) = k, v(t-1) = L \text{ and } e(t) = e_d).$$

Now we have an induced the Markov chain on the states $(C(t), v(t)\ )$. Let us denote the states of this Markov chain by $s_{ij}, i = 1, 2, \ldots m, j = 1, 2, \ldots n$. Thus state s_{ij} corresponds to concentration level C_i and ventilation level v_j.

Suppose we have as our air quality standard both requirement (2) and the requirement that the annual average concentration should be less than β units.

Let k_d be the cost of operating one day at the emission level e_d.

Then the problem of setting dynamic emission standards so as to minimize costs while meeting air quality standards can be shown to be the following linear programming problem [3]:

$$\text{Min} \sum_{i=1}^{m} \sum_{j=1}^{n} \sum_{d=1}^{q} k_d w_{ijd}$$

---

* In fact, any relation of the type $C(t) = f(C(t-1), u(t), v(t)\ )$ where f is an arbitrary function can be treated with the analysis outlined here.

subject to

$$-\sum_{d=1}^{q} w_{ijd} + \sum_{k=1}^{m} \sum_{L=1}^{n} \sum_{d=1}^{q} p\,(i,j\,|\,k,L,d)\,w_{kLd} = 0 \quad \text{for} \quad \begin{matrix} i = 1, 2, \ldots, m \\ j = 1, 2, \ldots, n \end{matrix}$$

$$\sum_{k=1}^{m} \sum_{L=1}^{n} \sum_{d=1}^{q} w_{kLd} = 1$$

$$\sum_{k \geq C_0} \sum_{L=1}^{n} \sum_{d=1}^{q} w_{kLd} \leq \alpha \tag{11}$$

$$\sum_{k=1}^{m} \sum_{L=1}^{n} \sum_{d=1}^{q} C_k\, w_{kLd} \leq \beta \tag{12}$$

$$w_{kLd} \geq 0 \quad \begin{matrix} k = 1, 2, \ldots, m \\ L = 1, 2, \ldots, n \\ d = 1, 2, \ldots, q \end{matrix}$$

where w_{ijd} are the decision variables which are solved for in the linear program. Physically w_{ijd} is the probability of setting emission level at e_d and being in state s_{ij}. Note that in general the solution will give us a randomized strategy as optimal. That is to say, our solution will say that when in state s_{ij} the emission level for the optimal dynamic strategy should be e_d for a certain percentage of the time (not necessarily 0 or 100%).

An important advantage of this approach is that the shadow prices on rows (11) and (12) explicitly give us the marginal costs of altering the air quality restrictions. Thus, it is possible to investigate the economic impact of raising or lowering the air quality standards.

To illustrate this, a small hypothetical model was solved on a computer.

For this example: $q = 2$, $m = 6$, $n = 4$

$$v_1 = 0,\, v_2 = 0.1,\, v_3 = 0.2,\, v_4 = 0.4$$

$$C_1 = 0,\, C_2 = 2,\, C_3 = 4,\, C_4 = 6,\, C_5 = 8,\, C_6 = 10$$

$$e_1 = 1,\, e_2 = 4,\, k_1 = 1,\, k_2 = 0$$

and the transition matrix $q(j\,|\,i)$ is given by the element in the ith row and jth column of the following matrix:

$$\begin{bmatrix} 0.5 & 0.3 & 0.2 & 0.0 \\ 0.2 & 0.5 & 0.2 & 0.1 \\ 0.0 & 0.3 & 0.5 & 0.2 \\ 0.0 & 0.2 & 0.3 & 0.5 \end{bmatrix}$$

The air quality requirements were given by $C_0 = 6$, $\alpha = 0.02$, $\beta = 3.0$.

The Table below shows the results of parametric runs on the model and illustrates the information these give on the cost effects of tightening or relaxing air quality standards.

Parametric analysis of a numerical Example: α = fraction of days during which critical concentration can be exceeded; β = maximum permissible average concentration.

Case No.	α	β	Minimum cost	Shadow prices on α row	on β row
1	0.020	3.0	0.1634	1.61	0
2	0.011	3.0	0.1779	1.61	0
3	0.005	3.0	0.1875	1.61	0
4	0.02	1.2	0.1634	1.61	0
5	0.02	0.6	0.3095	0	0.625

To obtain the cost for the static emission policy for comparison with the dynamic, we can use the following procedure.

1. Set $d = q$.

2. Solve $\sum_{k} \sum_{L} \Pi_{kL} p(i, j, L, d) = \Pi_{ij}$

$$\sum_{k} \sum_{L} \Pi_{kL} = 1$$

for Π_{ij} = steady-state probability of being in S_{ij} under static emission level e_d.

3. If $\sum_{k > C_0} \sum_{L} \Pi_{kL} \leqslant \alpha$ and $\sum_{k} \sum_{L} \Pi_{kL} C_k \leqslant \beta$ then quit. The value of d obtained is the optimal for a static policy. Otherwise reduce d by 1 unit and repeat step 2.

For our example problem the cost for the static policy is 1.

4   **Conclusion**

In this paper I have suggested that dynamic standards are more efficient than static standards for control of large pollutant sources. I have analyzed a simple model which provides some insight into the nature of dynamic standards and a more realistic model which would help in optimally setting dynamic standards. These models will also aid quantitative estimation of the advantages to be gained from dynamic over control policies.

One aspect of dynamic strategies which has implications for research and development of antipollution devices is worth noting. The dynamic strategy may render feasible techniques and apparatus for pollution control which have high operating costs. In such a case, if one computed the cost for a static emission standard it may be too exorbitant to impose on the polluter. However, the dynamic strategy may require only intermittent use of the expensive pollution-control equipment, and hence may reduce costs to an acceptable magnitude. Thus, research efforts could be justifiably directed towards developing pollution control technology even if it has high operating costs.

Acknowledgments

I would like to record my indebtedness to Prof. J. R. MAHONEY of the Harvard School of Public Health and Prof. J. D. C. LITTLE of the Massachusetts Institute of Technology for their help and encouragement. I would also like to express my thanks to the US Army Research Office (Durham) for funding my research work.

References

[1] National Primary and Secondary Ambient Air Standards, Federal Register *36*, 8186–8201 (30 April 1971).
[2] STEVEN J. MARCUS, *Mathematical Decision Models for Air Pollution Control Policies*, unpublished Ph. D. Thesis, Harvard University (January 1971).
[3] HARVEY M. WAGNER, *Principles of Operations Research* (Prentice-Hall, 1969).

A Method for the Optimal Location of Urban and Industrial Agglomerations

by STANISLAW CHRÓSCIEL and MACIEJ NOWICKI
Institute of Environmental Engineering, Warsaw Technical University,
Poland

Abstract

A new method for the optimal location of urban and industrial agglomerations, based on a 'pollution does factor', is presented. The 'pollution dose factor' allows the comparison of various pollutants in one and the same model and makes simple optimalization procedures possible. As an example several theoretical town models are compared from the air pollution point of view. Possible extensions of the model for regional studies are discussed.

1   Introduction

1.1   *Criteria for location of industrial plants*

Ever since its beginnings, industry has always been the most important town-creative element and the main stimulator for the demographic and territorial development of towns. But the development of the towns in the nineteenth century was a chaotic one, uncontrolled by the administrative authorities and without any urban concepts. This has caused the considerable deformation of the urban system which today is characterized by a mixture of industrial and urban agglomeration and by excessive population concentrations.

In the same period, the factors deciding the selection of a location for a new industrial plant were governed by a microeconomic approach on the part of the individual businessman. Such an approach was formalized by the first theorist of industrial regional planning, WEBER [1], according to whom the minimum costs of the transportation of raw materials, semimanufactured goods and final products is the main factor in determining the location of an industrial plant. Certain deviations from this approach, however, were caused by the minimum costs of labour or the advantages connected with the urban agglomeration.

In the twentieth century, along with technological progress and improvement of the transportation system, transportation costs decreased while the importance of sales and labour market as the principal factors in the location of an industrial plant increased [2, 3]. At the same time the protection of the natural environment, and, in particular, water and air protection, played a more and more important role in determining the location of an industrial plant. While in 1909 WEBER [1] wrote: 'Water is practically unlimited and therefore absolutely available in many regions', already by 1952 the United States Presidential Commission on Materials Policy [4] had declared that: 'About 1975 the availability of good water may become the most important element of the decision on the location of an industrial plant.' Thus, within a very short period of time, water has become a scarce resource in many regions.

The significance of air protection against pollution was increasing parallel to that of water protection, though with a certain delay which was caused by incomparably larger although admittedly *limited* resources of air. Due to the considerable increase in the concentration of toxic substances in industrial regions and to the world-wide activity for environmental protection, the problem of air protection has become in recent years one of the main factors influencing the location of new industrial plants.

1.2 *Disproportion in location of industry*

The traditional industrial centers, increasing spontaneously since the nineteenth century, represent nowadays regions of monstrous dimensions in which the natural environment has degenerated considerably and air pollution is an imminent threat to human health and life. Improvement of the living conditions in these areas requires now a strict control and evaluation of every new investment from the point of view of its indispensability and its influence on the environment. At the same time, the deglomeration of the most polluting industrial plants, the liquidation of some old ones and the modernization of the technology of those which are indispensable in the given areas must take place in order to reduce the emission of compounds into the atmosphere.

Parallel to those over-populated and excessively industrialized regions, there are in almost every country less developed regions which are characterized by a lower living standard and a surplus of labour force. Thus the main investment effort should be directed towards those areas. The achievements of Italy [5], Yugoslavia [6], or Poland [7] exemplify that the industrialization of these regions is the only way to achieve rapid activation, and that the industrial plants should be concentrated at certain carefully selected points which could be called the 'poles of industrial growth'. According to the United Nations Economic Commission for Europe [8], the construction of one or more industrial plants is indispensable in order to induce the required transformation of the less developed region. Obviously, the main criterion in the selection of poles for industrial growth, will be related to the economics of a region. At the very first stage of an industrial planning the principles of air protection should be taken into account in these areas to prevent repetition of the errors made in the traditional industrial regions. Therefore, it seems that the determination of the optimal regional relationship among the industrial, urban and commercial districts is the most important problem of regional planning in less developed regions. It should be emphasized that all theoretical conceptions of ideal towns presented in literature starting, with the concentric models by FRITSCH or HOWARD through various models of the linear towns by CARNIER, LE CORBUSIER DAHL, HILDERSEIMER or SCHÜRMANN to the models of the multicellular towns by WRIGHT or GLOEDEN do not take into account the importance of air protection when considering the influence of industry on an urban area, except certain attempts in HILDERSEIMER's model. The most important factor when analysing the location of the industrial district in relation to the town in these models is to ensure a functional communication system between residential areas, business areas, and the commercial center. Thus it seems necessary to create a systematic comparison of these models which from the point of view of atmospheric pollution should become indispensable for the rational design of a new town or for the planning of developmental directions of an already existing urban center.

Another important problem in regional planning is the determination of the optimal location for a new industrial plant so that the inavoidable pollution emission would be the lowest possible, not only for the town as a whole, but especially for certain important districts or buildings (for instance for health resorts, recreation areas, hospitals, etc.).

The above problems have not been solved satisfactorily. This paper attempts to present a possible solution for optimization of locations for industrial plants.

2 Applied Mathematical Model

2.1 *Pollution dose*

The method considering *various* kinds of air pollutants occurring simultaneously has not been elaborated up to now. We have attempted a first solution to this problem which has been presented in detail in [9–10].

The idea of this method is the following: every kind of air pollutant is represented by the coefficient k, the 'coefficient of pollution toxicity', which is inversely proportional to the permissible value of the mean annual concentration. It is assumed that $k = 1$ for SO_2 (reference pollutant).

For instance, if the permissble mean annual concentration for SO_2 amounts to 0.05 mg/m³ for one pollutant and 0.015 mg/m³ for another, the coefficient for this substance will be: $1.0.05/0.015 = 3.33$.

Expressed in general terms, the total 'pollution dose' of g substances is $=$

$$d = \sum_{t=1}^{t=g} \overline{S}_t\, k_t . \qquad (1)$$

$d =$ mean annual concentration calculated simultaneously for all kinds of air pollutants and weightened by the resp. toxity in relation to the SO_2 concentration ('pollution dose').

$\overline{S}_t =$ mean annual or long-term pollution concentration of pollutant t.

If an emission source e emits g air pollutants, its emission E_e in relation to SO_2 can be defined by analogy to (1):

$$E_e = \sum_{t=1}^{t=g} E_{et}\, k_t \qquad (2)$$

The pollution dose determined by (1) represents the cumulative influence of various kinds of pollutants but it cannot be identified with the synergetic influence of pollutants. The theoretic pollution dose for a given point P and for the emission source e is:

$$d_{ep} = \sum_{i=1}^{i=k} \sum_{b=1}^{b=a} S_{epib} \, v_{ib}$$

i = wind velocity groups used for the calculations
k = number of wind velocity ranges (c.f. wind rose in Fig. 2)
b = segments of the wind rose
a = number of wind direction segments (c.f. Fig. 2)
S_{epib} = momentary pollution concentration at the point P_p in sector b caused by the emission source E_e for the wind velocity
v_{ib} = frequency of the wind of velocity i in sector b

The value S_{epib} can be calculated on the basis of any theoretical formula (SUTTON, PASQUIL, etc.) but the phenomenon of the pollution atrophy, layers position of the temperature inversion, status of the atmospheric equilibrium, as well as other factors essential for the proper description of the air pollution status have to be considered.

The pollution dose is a very convenient measure to evaluate the air pollution degree because of several essential charcteristic features, namely:

a) It enables the evaluation of the cumulative effects of various kinds of pollutants.

b) It is an additive value, i.e. it can be calculated separately for each pollution source and also for any number of combined sources. This feature is especially important when analysing various variants of the emission sources location.

c) It permits consideration of elements existing in the mathematical model such as pollution quantity and emission conditions, atmospheric status, pollution atrophy, etc.

d) It enables the use of computers; the analysis can thus involve hundred- of emission sources and places. It permits analysis of homogeneous and heteros geneous line, area, and volume emission sources by replacing them with the adequate number of point sources.

All the above reasons have caused the pollution dose to be the accepted measure in this paper for the evaluation of various possibilities of industrial plant locations and their influence on urban areas.

2.2 *Possibilities for application of the pollution dose for studies on air imminence status*

The pollution dose for one kind of pollutant represents its mean concentration for a long period of time, usually for one year. The values of the permissible mean annual pollution concentration are standardized in certain countries so the pollution dose for one kind of pollutant can be compared with them. When various kinds of pollution exist simultaneously the comparison of the pollution dose with the standardized values is of an approximate character but can be a criterion of general imminence to the area.

When analysing the influence of air pollution on forests [9, 10, 14] the effect of the pollution dose on vegetation can be used as a measurement of the influence of pollution on the forests. It has been stated that as the dose value amounted to about 0.40 mg/m³, the amount obtained by measurements and by theoretical studies, certain damage to the conifers were noticed. The consequence of these studies is a methodology for calculation of the repartition of emission sources in damage to forests [13]; it is today connected with the resolution of the Polish Council of Ministers on the responsibility of industrial plants for damages to the forests caused by air pollution. The advantages of the air pollution dose have resulted in its frequent application in various studies carried out by the Institute of Environmental Engineering at the Technical University of Warsaw. The pollution dose has been applied for various studies:

to determine the pollution background for large areas,

to correct the mathematical models describing the status of air pollution,

to evaluate the general pollution concentrations over large areas.

It can be concluded from the above that the 'pollution dose' concept may be regarded as a quite comprehensive and workable system for the evaluation of a broad number of air pollution problems.

2.3 *The application of the pollution dose model for the optimal location of industrial plants and urban agglomerations*

When a new industrial plant is to be built in a different location, air pollution must be an important criterion in the decision. The additional pollution doses due to the plant (or plants) may be integrated over certain designated regions (recreation areas, hospitals, etc.) or the whole city. By running the model calculation for the various possible alternatives, the optimal location can be determined. The importance of clean air in certain specific areas should be taken into account if calculation of the whole area or city is undertaken; this is possible through the introduction of a factor C_p, which weighs the importance of population density, recreation requirements, etc. in every area P. The total 'effect' of the industry with plants at new possible locations in a city of '1' inhabitants, expressed as the average weigthened pollution doses per citizen is thus

$$D = \frac{1}{l} \sum_{e=1}^{e=m} \sum_{p=1}^{p=l} d_{ep} \, C_p$$

At the level determined for emission conditions and at that determined for meteorological parameters the dose d_{ep} depends only on the location relationship between the emission sources and the area of observation P.

This model is not only valuable in determining the location for new plants in existing areas, but also in establishing new recreation and living zones for a city with established industrial zones.

3   Example: Comparison of Theoretical Town Models

In addition to real-world applications, the method described above may also be used for theoretical studies of the optimal town model. Thus it represents a valuable tool for including the air pollution aspect in these theories.

The concentric town models by FRITSCH and HOWARD, the linear models by MILUTIN and DAHL, the satellites by SCHÜRMANN discussed extensively in literature will now be analyzed from the air pollution point of view. It is necessary to relate these models to some initial assumptions:

Population: According to ISARD [15], MALISZ [16], and KAWALEC [17], the most adequate municipal unit for industry location is the town of 100,000–200,000 inhabitants. We have thus assumed a town of 200,000 inhabitants for this comparison.

Town area: An average European city has a population density of 10,000–20,000 inhabitants per km². We have assumed 10.000 inhabitants, and will use for this general model comparison an average distribution of population density: the town area is thus 20 km².

Pollution sources: Every model city has 8 large industrial plants emitting harmful pollutants into the air. The emission value of each source, expressed in relation to SO_2 as explained in chapter 2 has been assumed to be 100 g/sec and the velocity of the gasses at the emission outlet 25 m/sec, at a temperature of 400°K and a chimney height of 100 m. The industrial plants are located at the same distance from the urban district in all models.

Meteorological parameters: The wind direction pattern is the same for all towns with prevailing south-western and western winds (c.f. windrose SUTTONS equation has been used for all concentration calculations. The meteorological exponent and the atmospheric diffusion coefficients are those commonly used in Poland. The studies have concentrated only on the comparison of the optimal location in each model.

Result: The pollution dose distribution and mean dose over the whole town are shown in Figure 1 through Figure 11. On the basis of the presented calculations it can be concluded that the best conditions exist in the towns with linear character and proper orientation in relation to prevailing wind directions. This conclusion is an additional argument for the town-planners who have been postulating for several years to give up the concentric models because of several other advantages characterizing the linear ones (functionality of the transportation system, connection with the surrounding natural environment, etc.). Nevertheless, any variation of the location of the town should be analyzed each time in detail because in a case of an inappropriate regional relationship between industry and urban districts, all advantages of the linear model cannot be gained. The example of an inappropriate location of industry is illustrated in Figure 5. The change of the industrial plants to the other side of the town has caused an increase of the mean dose by 35 per cent (from 0.0708 to 0.0975 mg/m³) and made this alternative variant the worst one among all models under consideration, including the concentric ones. Also thorough analysis of such equivalent models as presented in Figure 2 and 3 permits one to draw the conclusion that the NW–SE structure of the town is better because the air pollution loads are nearly equal over the whole area of the town (the differences between the loads on the particular districts do not exceed 35 per cent whereas these differences can be 115 per cent if a NS orientation of the town is assumed.

Another conclusion drawn from the calculations is the equivalence from the point of view of air pollution of three analysed concentric models with the commercial center in the middle and the urban and industrial districts surrounding it. Especially disadvantagous conditions exist in the satellite town by SCHÜRMANN (Fig. 10). There are some districts of a very small pollution dose and nearby others of a very high one amounting to 0.14 mg/m³; the difference between the mean dose on the individual areas is about 240 per cent.

The calculations have confirmed the inappropriateness of the traditional settlement structure, with the shape of a circle or a square with the industrial districts on its edges. The mean pollution dose is in this case 75 per cent higher than that in the linear town.

4   Final Conclusions

4.1   *Possibilities of the application of the mathematical model*

The pollution dose is a convenient instrument for the evaluation of air protection policies for the urban and industrial agglomeration. The pollution dose calculated according to the scheme presented above can serve:

to optimize the shape and location of a new town and the location of its industrial plants,

to determine the optimal air protection policy which should be implied in existing towns,

to evaluate the impact of an individual industrial plants of the town, or certain districts and buildings,

to optimize the location of a new industrial plant in relation to the existing urban and industrial agglomeration.

Any greater regional structure can also be analysed by the method.

4.2   *Perspectives for development of method*

The determination of the 'air pollution dose' poses today still several difficulties which reduce its universality and precision. The methods for determining the degree of toxicity for a given compound on an absolute scale and also on a relative scale to SO_2 toxicity are still unprecise. Up to now there exist not much information on the synergetic effects of various toxic

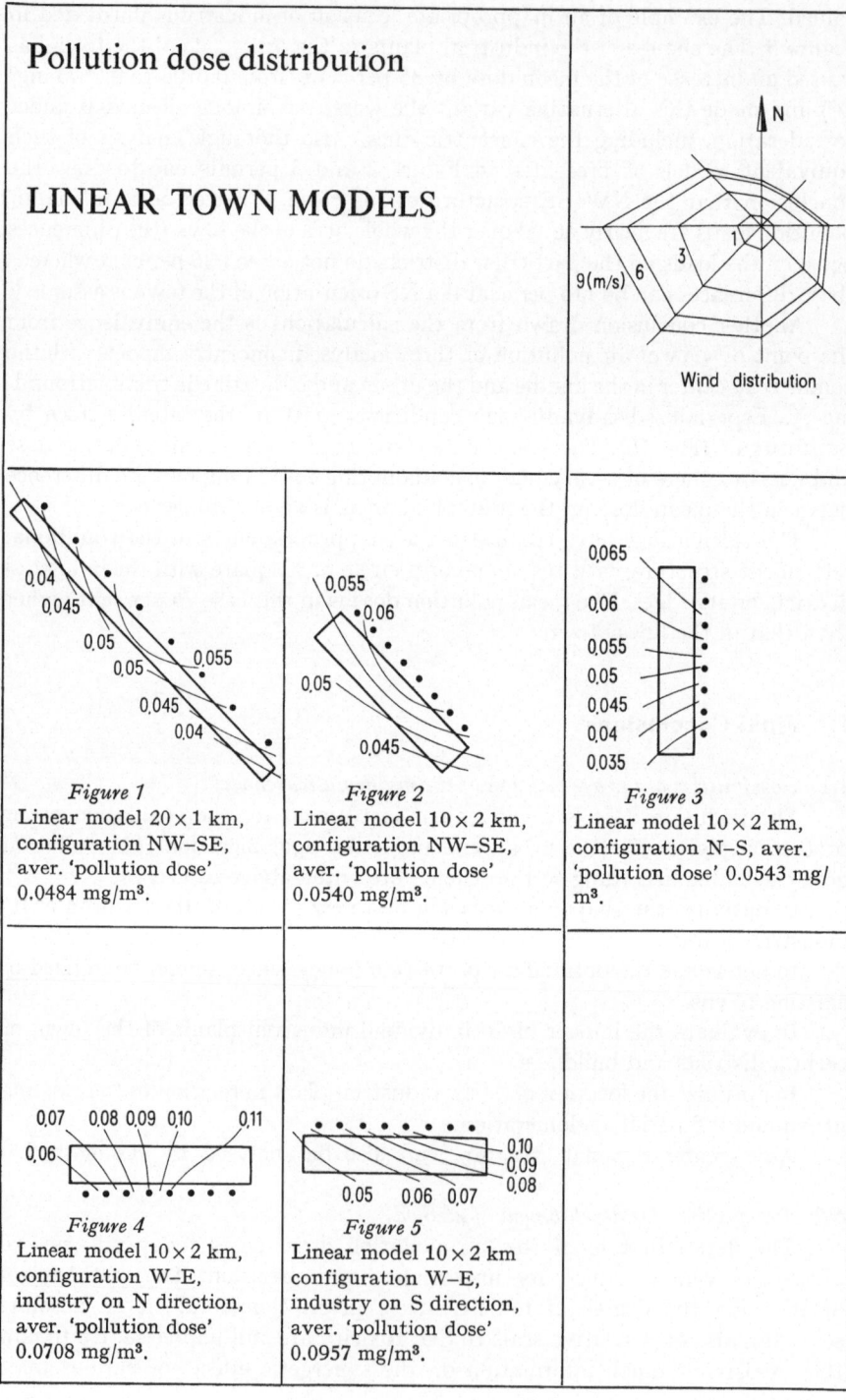

Pollution dose distribution

LINEAR TOWN MODELS

Wind distribution

Figure 1
Linear model 20 × 1 km,
configuration NW–SE,
aver. 'pollution dose'
0.0484 mg/m³.

Figure 2
Linear model 10 × 2 km,
configuration NW–SE,
aver. 'pollution dose'
0.0540 mg/m³.

Figure 3
Linear model 10 × 2 km,
configuration N–S, aver.
'pollution dose' 0.0543 mg/
m³.

Figure 4
Linear model 10 × 2 km,
configuration W–E,
industry on N direction,
aver. 'pollution dose'
0.0708 mg/m³.

Figure 5
Linear model 10 × 2 km
configuration W–E,
industry on S direction,
aver. 'pollution dose'
0.0957 mg/m³.

Pollution dose distribution

CIRCULAR TOWN MODELS

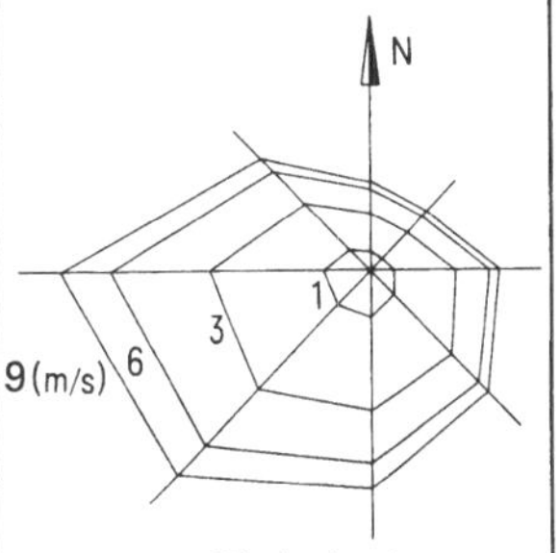

Wind distribution

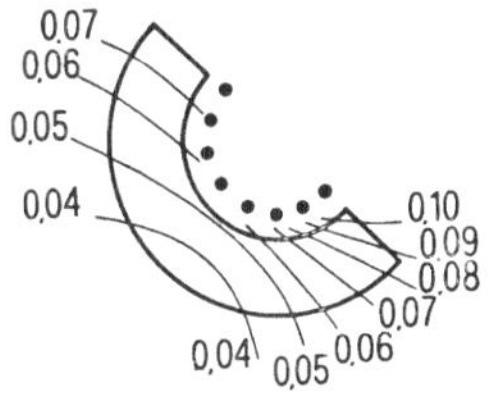

Figure 6
Semicolar model 10 × 2 km,
configuration NW–SE,
aver. 'pollution dose'
0.0610 mg/m³.

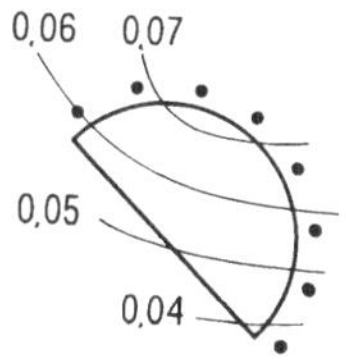

Figure 7
FRITSCH, semicircle
model, configuration
NW–SE, aver. 'pollution
dose' 0.0561 mg/m³.

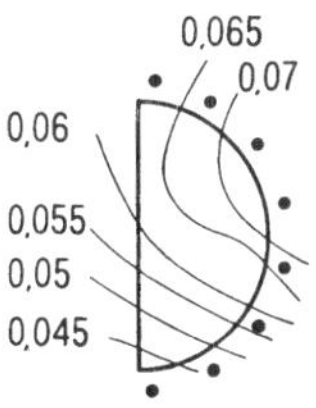

Figure 8
FRITSCH, semicircle model,
configuration N–S,
aver. 'pollution dose'
0.0591 mg/m³.

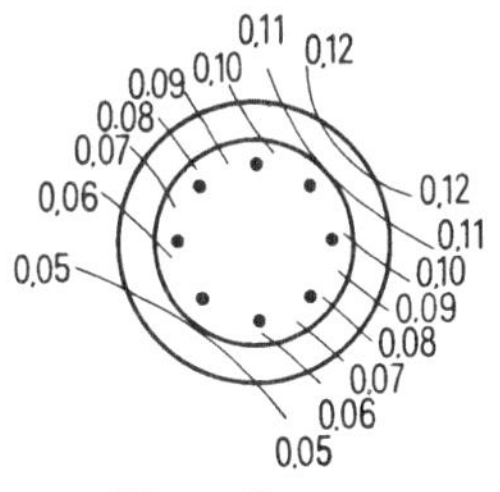

Figure 9
Collar model, aver.
'pollution dose'
0.0787 mg/m³.

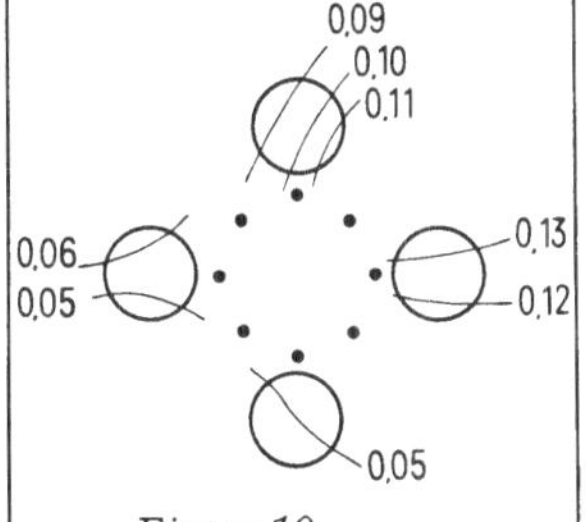

Figure 10
SCHÜRMANN, satellite model
aver 'pollution dose'
0.0810 mg/m³.

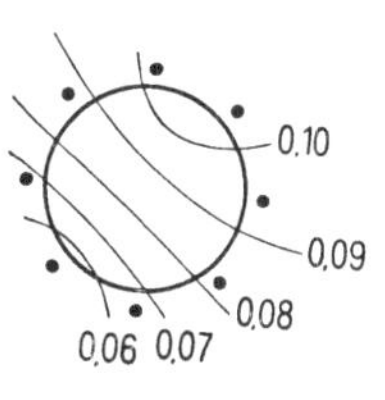

Figure 11
HOWARD, concentric model,
aver. 'pollution dose'
0.0850 mg/m³.

compounds, nor is there a quantitative description of the strophy of the particular pollutants concentration over a period of time.

The imperfection of the mathematical formula of SUTTON or PASQUILL, and in particular the difficulty in a proper selection the meteorological exponent and the atmospheric diffusion coefficients for a climatic and topographic condition represent other drawbacks, which are common to all mathematical models involving the status of the atmosphere in the vicinity of the pollution emission sources.

These difficulties can be solved only by experimental studies in physics and chemistry of the atmosphere, meteorology and toxicology. The progress in these scientific disciplines furthers the precision of the mathematical models and consequently leads to a more accurate determination of the pollution dose for an analysed area.

Nevertheless, it is necessary to emphasize that the pollution dose need not and even should not be the only criteria for analysing the air polltuion influence on the urban and natural environment. It would be most appropriate to include all values of permissible concentrations for the determined toxic substances; the studies will then, however, become very complex and expensive. The pollution dose is an attempt to find a simple and rapid method for carrying out such analyses.

References

[1] A. WEBER, *Über den Standard der Industrien* (Tübingen 1909).
[2] W. F. LUTTRELL, *Factory Location and Industrial Movement* (London 1962), p. 48.
[3] International Information Central for Local Credit, Government Measures for the Promotion of Regional Economic Development (The Hague 1964).
[4] Resources for Freedom, Vol. 1 (Washington D.C. 1952), p. 50.
[5] Expériences et perspectives d'interventions variant selon les régions et d'après la planification territoriale en Italie. (Rome 1964), p. 5–6.
[6] Le développement des régions sous-développées de Yougoslavie (Geneva 1964), p. 6.
[7] K. SECOMSKI, Ekonomista *4*, 735 (1964).
[8] Economic Commission for Europe U.N., Criteria for Location of Industrial Plants (New York 1967), p. 62.
[9] S. CHRÓŚCIEL and J. JUDA, *Metodyka oceny stopnia zagrozenia kompleksów leśnych przez określone źródla emisji*, Methodology for Evaluation of Imminence Degree to Forests by Determined Emission Sources (Kraków 1969).
[10] J. JUDA, S. CHRÓŚCIEL, JĘDRZEJOWSKI and M. NOWICKI, *Wpływ zanieczyszczeń emitowanych do atmosfery na róslinność Nadleśnictwa Olkusz*, Influence of Pollutants Emitted into Atmosphere on Vegetation of Olkusz Forest Inspectorate (Warszawa 1970).
[11] J. JUDA, J. PALUCH and S. CHRÓŚCIEL, *Methodik der Beurteilung des Gefährdungsgrades von Waldkomplexen auf Grund der Bestimmung der Emmissionsquelle*, VII. Internationale Arbeitstagung Forstlicher Rauchschadensachverständiger, Essen 1970.
[12] J. JUDA, JĘDRZEJOWSKI and M. NOWICKI, *Methodology of Atmospheric Air Pollution Measurements* (Springlied, Virginia 22151, 1971).
[13] Wytyczne Ministra Leśnictwa i Przemysłu Drzewnego Directives of the Minister of Forestry and Timber Industry of 19th Sept. 1970. (Kraków 1971).

[14] J. Juda, S. Chrościel, J. Jędrzejowski and M. Nowicki, *Ocena ilościowa wpływu przemysłowych zanieczyszczeń powietrza na Narodowy Park Ojcowski*, Quantitative Evaluation of influence of Industrial Air Pollution on Ojcow National Park (Kraków 1972).

[15] W. Isard, *Location and Space Economy* (New York 1956), p. 182–188.

[16] B. Malisz, *Zarys teorii kształtowania układów osadniczych*, Outline of Theory of Settlement Structures Formation (Warszawa 1966).

[17] B. Kawalec, *Biuletyn Komitetu Przestrzennego Zagospodarowania Kraju*, Bulletin of Regional Development Committee, Polish Academy of Sciences 6/15/1962.

[18] Niveaux optimes des villes – Essai de définition d'après l'analyse des structures urbaines (Lille 1959).

About the Authors

STEPHEN H. SCHNEIDER, is presently at the National Center for Atmospheric Research in Boulder, Colorado. His doctorate thesis at the Mechanical Engineering Department of Columbia University dealt with computational methods in fluid and plasma dynamics. Working at the Goddard Institute for Space Studies, Dr. SCHNEIDER became interested in fluid dynamics and radiative transfer, and together with Dr. S. I. RASOOL a model was developed to evaluate the effects on atmospheric particles and CO_2 on the radiation balance. In connection with the 'International Study of Man's Impact on Climate' (SMIC), held in Sweden in 1971, Dr. SCHNEIDER was responsible for writing several sections of the SMIC Report.

RODERICK W. SHAW is a research chemist in the field of air pollution meteorlogy at the Atmospheric Environment Service (AES), Toronto, Canada. He earned his BS at Queens University, Kingston, Ontario and his M.A. in physics at the University of Toronto. He was awarded his Ph. D. from McGill University on the subject of *Movement, Morphology and Circulation of Thunderstorms*. Dr. SHAW's research convered such subjects as *The Effect of Basin Precipitation on Lake Levels* and *Environmental Errors in the Use of the Airborne Radiation Thermometer*.

DOUGLAS M. WHELPDALE is a research scientist at the Atmospheric Environment Service (AES), Toronto, Canada. He graduated in mathematics and physics from the University of Toronto in 1964 and earned his M.S. in 1967 on the subject of *Factors Affecting the Coalescence of Cooling Water Drops*. In 1970 he received his Ph. D. for a thesis on *Raindrop Growth by Coalescence*.

SATU HUTTUNEN is working in the Department of Botany at the University of Oulu, Finland. His research covers problems of plant ecology and eco-physiology with special emphasis on their effect on forest vegetation, trees and dwarf shrubs. He has published extensive works on methodological investigations of air pollution under Northern European Conditions.

PAVLE TODOROVIĆ is the head of the Section for Traffic Medicine in the Institute of Public Health of the SR of Serbia, Yugoslavia. He graduated from the University of Belgrade in 1963 and earned his M.S. in 1969 and his M.D. in 1972. He has published several scientific and professional papers on the influence of environment on man and drivers. Dr. TODOROVIĆ's main interest centers around the influence of environmental factors on traffic accidents, and the physical and chemical factors influencing the driving capability.

S. J. FORMOSINHO is assistant professor at the University of Coimbra Portugal. He graduated with honors in 1964 followed by an assignment as lecturer at Coimbra University. He received his Ph. D. from the University of London in 1971. Dr. FORMOSINHO is presently in charge of chemical research at the Air Pollution Center of the Geophysical Institute leading a group of scientists in the petrochemical field. His main interest is photochemistry of aromatic compounds.

AUGUSTO C. CARDOSO is working with Dr. FORMOSINHO in the Air Pollution Center of the Geophysical Institute, Coimbra, Portugal, specializing in chemical research on air pollution.

JOSEF RIEDERER heads the Scientific Laboratory of the Geology Department of the Bavarian Art Galleries (Doerner Institute). He studied geology at the University of Munich, Germany, where he received his Dr. rer. nat. in 1964. He then served as assistant professor for three years after which he specialized in his present field. Dr. RIEDERER is responsible for the examination and conservation of art made of stone, glass, ceramics and metals and is a member of the Institute for Conservation (IIC) and the International Council of Museums (ICOM).

ERIC HOEDEL, joined after his doctorate the faculty of the Technische Hochschule, Darmstadt, Western Germany .Dr. HOEDEL is presently engaged in teaching and research on problems of environmental economics at the 'Institute for Macro- and Structural Planning'.

CHRISTOPH ANDREAS UHLIG is presently assistant at the University of Zürich, Switzerland. He graduated in engineering from the Oscar von Miller Polytechnikum, Munich, Western Germany. He later studied economics at the Ludwig-Maximillian University in Munich and received his Masters degree (lic. oec. publ.) at the University of Zürich, where he is presently defending his doctoral dissertation. His scientific interests are mainly concerned with environmental economics.

PETER H. SAND is presently Legal Officer with the UN Food and Agriculture Organization (FAO) in Rome, Italy. He was formerly associate professor of Law at McGill University, Montreal, Canada, and a visiting Fellow in Environmental Studies at the Smithsonian Institute in Washington. He is author of the FAO study on *Legal Systems for Environment Protection* and numerous papers in law reviews.

MICHAEL G. MCGARRY is associate professor of Environmental Engineering at the University of Western Ontario and presently also associate professor at the Asian Institute of Technology, Bangkok, Thailand, where he specializes in environmental aspects of developing countries. Dr. McGarry studied civil and sanitary engineering at the University of Alberta and received hisPh. Dr. from the University of New South Wales, Canada, in the field of Public Health Engineering. Dr. McGarry has also been an active consultant and advisor fo various governments.

J. I. Steinfeld is associate professor at Massachusetts Institute of Technology. He earned his Ph. D. from Harvard University in physical chemistry in 1965. Dr. Steinfeld was an Alfred P. Sloan Research Fellow in 1969–71 and presently holds a J. S. Guggenheim Memorial Fellowship. His research covers fields of molecular spectroscopy, molecular energy transfer, application of lasers to chemistry and his latest work deals with optical detection of atmospheric pollution. Dr. Steinfeld is a member of the American Physical Society, the American Association for the Advancement of Science and the Federation of American Scientists.

B. D. Green is a research assistant working with professor Steinfeld on laser ranging systems. He graduated from the University of Pensylvania with honors in chemistry and is presently working at M.I.T. towards his Ph. D. in physical chemistry.

Jayshree Manohar Deshpande of India received her M.S. degree in chemistry from Nagpur University, Nagpur, India. Miss Deshpande has been conducting scientific research on various environmental problems since 1964 at the Central Public Health Engineering Research Institute at Nagpur. She has been a visiting scientist at the University of Gothenburg, Sweden for one year and has now returned to India to continue her work on environmental problems.

Dennis Thompson is a Research Fellow in the Department of Physical Chemistry, Leeds University, England. He graduated in Physics and Chemistry at the University of Sheffield in 1966. Under the direction of Professor J. M. Beer he carried out research in nitric acid formation at the Department of Fuel Technology and Chemical Engineering and obtained his Ph. D. in 1971. His present research is concerned with estimation of heat loss rates during exothermic reactions in stirred systems.

Nitin R. Patel is presently working toward his Ph. D. in Operation Research at the Massachusetts Institute of Technology. He graduated from the University of Bombay in 1962 and continued his studies at M.I.T. where he received a M.S. degree in both Electrical Engineering and in Management. He then worked as a consultant in both USA and India for six years before returning to M.I.T. for his doctorate.

STANISLAW CHRÓŚCIEL is a lecturer at the Institue of Sanitary Engineering, Warsaw Techincal University, Warsaw, Poland, from where he also received his M.S. degree in Mechanical Engineering. He has since worked under Professor B. STEFANOWSKI and Professor B. STANISZEWSKI on thermal processes. He received his Doctor of Engineering degree on research in high frequency temperatures of gas streams and from 1966 he has been involved in work on air pollution, specializing in aerosol filtration.

MACIEJ NOWICKI is a researcher at the Warsaw Techical University in the Institute of Environmental Engineering. He received his M.S. in Sanitary Engineering in 1964 and continued his studies at the Polish Academy of Sciences. He received his Doctor of Engineering degree on the thesis of aerosol filtration of fibrous filters. Dr. NOWICKI's work also covers pollution dissemination in the atmosphere and measurements of industrial dust parameters.

JAN-OLAF WILLUMS is presently a research assistant at the Massachusetts Institute of Technology, Cambridge, USA. He graduated with a M.S. in Mechanical and Industrial Engineering from the Swiss Federal Institute of Technology in 1971. In 1969 he was an exchange student at Waseda University, Tokyo, Japan, and in 1970 at Stanford University. In 1972 he was awarded a Fulbright Fellowship and the Norwegian-American Fellowship for his doctoral studies on interdisciplinary problems of the Ocean Environment at M.I.T.